The Genesis of the Ores of Tonopah Nevada

by the U.S. Dept. of Geological Survey

with an introduction by Kerby Jackson

Introduction

It has been almost a century since the Department of Interior released it's important publication "The Genesis of the Ores of Tonopah, Nevada". First released in 1918, this important volume has been out of print and has been unavailable to the mining community since those days, with the exception of expensive original collector's copies and poorly produced digital editions.

It has often been said that "*gold is where you find it*", but even beginning prospectors understand that their chances for finding something of value in the earth or in the streams of the Golden West are dramatically increased by going back to those places where gold and other minerals were once mined by our forerunners. Despite this, much of the contemporary information on local mining history that is currently available is mostly a result of mere local folklore and persistent rumors of major strikes, the details and facts of which, have long been distorted. Long gone are the old timers and with them, the days of first hand knowledge of the mines of the area and how they operated. Also long gone are most of their notes, their assay reports, their mine maps and personal scrapbooks, along with most of the surveys and reports that were performed for them by private and government geologists. Even published books such as this one are often retired to the local landfill or backyard burn pile by the descendents of those old timers and disappear at an alarming rate. Despite the fact that we live in the so-called "Information Age" where information is supposedly only the push of a button on a keyboard away, true insight into mining properties remains illusive and hard to come by, even to those of us who seek out this sort of information as if our lives depend upon it. Without this type of information readily available to the average independent miner, there is little hope that our metal mining industry will ever recover.

This important volume and others like it, are being presented in their entirety again, in the hope that the average prospector will no longer stumble through the overgrown hills and the tailing strewn creeks without being well informed enough to have a chance to succeed at his ventures.

Kerby Jackson
Josephine County, Oregon
May 2015

CONTENTS.

3

ILLUSTRATIONS.

4

THE GENESIS OF THE ORES AT TONOPAH, NEVADA.

By Edson S. Bastin and Francis B. Laney.

INTRODUCTION.

In the spring of 1915, while engaged in a study of silver enrichment for the United States Geological Survey, Mr. Bastin spent a month at Tonopah, Nev., in studying the ores and their mode of occurrence. At about the same time Mr. Laney was directed by the United States Bureau of Mines to make collections of Tonopah ores and to study their mineralogy in connection with the researches of that Bureau on their metallurgic treatment. To avoid duplication of effort and to secure the added fruitfulness which should follow the concentration of more than one mind on a scientific problem, it was decided to ccmbine these efforts in so far as the two fields of research overlapped. The paper accordingly presents the results of informal cooperation between the Geological Survey and the Bureau of Mines.

The investigation was intended to supplement the important work of Spurr[1] and Burgess[2] by applying to the ores methods of microscopic study which were not in general use by economic geologists at the time their reports were prepared, but which, in other districts, have proved of material assistance in the interpretation of ore genesis.

For the structural observations and the discussion of the mine waters the senior author is responsible; for the detailed mineralogic and microscopic studies both authors are responsible. They are indebted to Mr. Chase Palmer, of the United States Geological Survey, for careful analyses of samples of the mine waters and for chemical and mineralogic studies, and to Messrs. F. L. Ransome and Adolph Knopf, also of the Survey, for valuable criticisms and suggestions.

The mine operators and others in Tonopah gave generously of their time and their knowledge and contributed many choice specimens in furtherance of the work. To them the writers' sincere thanks are offered.

THE TONOPAH DISTRICT.

The Tonopah mining district is too well known to require an extended description. From the discovery of its ores in 1900, its output has increased year by year until in 1915 its production of silver was exceeded in the United States only by that of Butte. Its metal output since 1904 is shown in the following table, compiled by .V. C. Heikes,[3] of the Geological Survey:

Gold, silver, copper, and lead produced in Tonopah district, Nev., 1904–1916.

Year.	Number of producers.	Ore.	Gold.	Silver.	Copper.	Lead.	Total value.
		Short tons.		Fine ounces.	Pounds.	Pounds.	
1904	10	22, 703	$386, 526	2, 119, 942			$1, 594, 893
1905	12	91, 651	1, 206, 345	5, 369, 439			4, 449, 486
1906	9	106, 491	1, 304, 677	5, 697, 928			5, 122, 289
1907	10	214, 608	1, 183, 628	5, 370, 891	5, 939	195, 508	4, 739, 966
1908	13	273, 176	1, 624, 491	7, 172, 396			5, 425, 861
1909	9	278, 743	1, 400, 361	7, 872, 967	1, 784	1, 488	5, 494, 600
1910	18	365, 139	2, 303, 702	10, 422, 869	942	6, 902	7, 932, 475
1911	19	404, 375	2, 366, 495	10, 868, 268		14	8, 126, 677
1912	17	479, 421	2, 223, 878	10, 144, 987	14	700	8, 463, 079
1913	32	574, 542	2, 613, 843	11, 563, 437	1, 150	9, 001	9, 598, 733
1914	23	531, 278	2, 648, 833	11, 388, 452	2, 284	924	8, 946, 987
1915	22	516, 337	2, 228, 983	10, 171, 374			7, 385, 870
1916	21	455, 140	1, 941, 441	8, 734, 726			7, 688, 891
		4, 313, 604	23, 433, 203	106, 897, 676	12, 113	214, 523	84, 969. 807

[1] Spurr, J. E., Geology of the Tonopah mining district, Nev.: U. S. Geol. Survey Prof. Paper 42, 1905; Geology and ore deposits at Tonopah, Nev.: Econ. Geology, vol. 10, pp. 713–769, 1915.
[2] Burgess, J. A., The geology of the producing part of the Tonopah mining district: Econ. Geology, vol. 4, pp. 681–712, 1909.
[3] Heikes, V. C., U. S. Geol. Survey Mineral Resources, 1914, pt. 1, p. 702, 1915; idem, 1915, pt. 1, p. 647, 1916.

The copper and lead came from properties closely adjoining the district. During the years represented in the table dividends were paid amounting to $21,000,000.

The average ratio of the gold to the silver in ounces recovered during this period was about 1: 98.

The total average recovery value per ton of the ore produced in 1913 was $16.71, in 1914 $16.84, and in 1915 $14.30. Most of the ore mined is treated by cyanidation, with or without concentration. It is stated that in certain plants ore assaying as low as $8 a ton can under certain circumstances be profitably treated.

STRATIGRAPHY.

It was not the writers' purpose in visiting Tonopah to review the work of Spurr and of Burgess on the age and mutual relations of the rock formations of the district, nor would this have been possible in the short time available for field work, as will be readily appreciated by anyone familiar with the remarkable complexity of the geology. The divergent views of Spurr and Burgess have been summarized with unusual clearness by Augustus Locke.[1] Locke's own views are so nearly in harmony with those of Burgess that it appears unnecessary to present them separately. The following brief summary of the major stratigraphic features of the district will assist in understanding the data on the genesis of the ores presented in this paper.

The Tonopah district is underlain by a thick series of rocks that are products of volcanic activity and are believed to be of Tertiary age. Broadly classified, these rocks consist of a lower series, mostly rhyolitic, which is overlain by two andesites, distinguished as "earlier" (now termed Mizpah trachyte by Spurr) and "later (now termed Midway) andesite." The "later andesite" is overlain

Geologic formations at Tonopah, Nev.

Origin and relative age according to Spurr.	Approximate stratigraphic succession.	Origin and relative age according to Burgess.
11. Intrusive as volcanic necks.	Brougher dacite and Oddie rhyolite.	9. Surface flows and the conduits through which the lavas rose.
10. Surface flow.	Basalt.	8. Surface flow.
8b. Intrusive.	Tonopah rhyolite of Spurr in northern part of area.	6b. Surface flows.
9. Lake deposits of volcanic material.	Siebert tuff.	7. Lake deposits of volcanic material.
8a. Intrusive, surface flows and tuffs.	Tonopah rhyolite of Spurr in southern part of area.	6a. "Surface flows of rhyolites, and breccias." Not separately considered.
7. Tuffs and mud flows.	Fraction dacite breccia.	
6. Volcanic necks and possibly some surface flows.	Heller dacite.	
5. Surface flow.	Midway andesite or "later andesite."	5. Surface flow.
1b. Surface flow.	Mizpah trachyte or "earlier andesite."	4. Surface flows.
4. Intrusive sheet.	West End rhyolite or "upper rhyolite."	3. Surface flows.
3. Intrusive sheet with many inclusions.	Montana breccia.	Not specifically mentioned but evidently classed as surface volcanic rocks. (See Burgess, J. A., op. cit., p. 682.)
1a. Basal phase of Mizpah trachyte flow.	Glassy trachyte.	
2. Flat-lying intrusive.	Sandgrass or "calcitic" andesite.	2. Surface flow.
8c. Intrusive with "autobrecciation."	Lower rhyolite.	1. Complex of tuffs, breccias, and flows.

[1] The geology of the Tonopah mining district: Am. Inst. Min. Eng. Trans., vol. 43, pp. 157-166, 1912.

by rhyolite, lacustrine tuffs, and thin flows of basalt. In spite of complicated faulting most of the volcanic formations are rather flat-lying. The middle column of the table on page 8 shows the formations that have been recognized, arranged in the normal order of superposition. In the left-hand column is given Spurr's latest interpretation of their origin and age relations, and in the right-hand column Burgess's interpretation. The numbers indicate the relative ages under the two interpretations; formations that are regarded as contemporaneous are given the same number but differentiated by letter, as 8a, 8b. The two interpretations differ materially, most of the rocks which Spurr regards as flat-lying intrusive rocks being interpreted by Burgess as flows poured out upon ancient surfaces.

The decision between these two interpretations must be left to future students of Tonopah geology. Successful refutation of the detailed work of Spurr and his associates must be founded on observations comparable with theirs in detail and comprehensiveness.

RELATIONS OF ORES TO INCLOSING ROCKS.

The bulk of the metal production of the district has come from ore bodies lying wholly within the Mizpah trachyte ("earlier andesite"). Within recent years, however, as the result of deeper development and underground exploration of new territory, an increasing percentage of the output has come from ore bodies bounded on one or both sides by other formations. Many of the veins, for example, whose upper portions are wholly in the Mizpah trachyte have at greater depth one or both walls of West End rhyolite. Spurr[1] believes that the productive veins are of several ages and classifies them as first-period and second-period veins, the latter with four subclasses. He recognizes also third-period veins, which are not productive. The veins of the first period he believes are older than the Montana breccia and West End rhyolite; those of the second period are younger than these formations; and both sets are older than the "cap rock" or Midway andesite.

The writers had little opportunity to study in the field the structural evidence on which Spurr bases his classification. Locke[2] says: "There is no mineralogical distinction whatever to be made between many veins which, according to the hypothesis [Spurr's], should belong to different periods." The writers' observations tend to indicate at least the approximate correctness of Locke's statement. The principal productive veins of the region show notable similarity in the minerals present and in the mutual relations of these minerals. The ore of the Belmont vein, for example, which Spurr classes as a first-period vein, is practically identical in mineral composition with that of the MacNamara and Fraction veins, which he classes as second-period veins. Mineralogic differences between veins of supposed different ages are commonly no greater than may be observed in different parts of one continuous vein. In most of the "first-period" veins there is little evidence of fracturing of the ore and filling of the fractures by later ore except that resulting from downward enrichment. Upon the sole basis of the mineral composition and texture of the primary ore there would be little reason for the impartial geologic observer to regard the principal productive veins as of more than one age. Mineralogic similarity can not, however, be taken as proof of contemporaneity, for in a region where there has been recurrent volcanic activity it is possible if not probable that similar mineralization might take place at more than one period. Structural observations must furnish the final test.

In the descriptions here given the ores of the principal productive veins are described as a unit but without implication of precise equality in age. The wolframite of the Belmont vein is an exception, for it is shown by textural relations to be somewhat later than the silver ore with which it is associated.

As emphasized by Spurr[3] the ores of Tonopah are in the main replacement deposits along sheetlike zones of fracturing. They are therefore typical replacement veins. Every gradation may be traced from heavy sulphide ore into slightly replaced wall rock. Ore-cemented breccias, true crustification, comb structure, and other features characteristic of

[1] Spurr, J. E., Geology and ore deposits at Tonopah, Nev.: Econ. Geology, vol. 10, pp. 751-760, 1915.

[2] Locke, Augustus, The geology of the Tonopah mining district: Am. Inst. Min. Eng. Trans., vol. 43, p. 164, 1912.

[3] Spurr, J. E., Geology of the Tonopah mining district, Nev.: U. S. Geol. Survey Prof. Paper 42, p. 84, 1905.

the filling of open spaces are rare or developed only on a small scale. Banding somewhat resembling true crustification, which is locally conspicuous, is possibly to be explained by diffusion [1] during replacement.

Most of the veins dip steeply, but some, like the West End and MacNamara, are only moderately inclined.

DIVERSITY IN MINERALIZATION.

Whatever may be the differences in age among the veins of Tonopah, it is certain at least that most of the ores represent the net result of several successive geologic processes. A study of their texture and mineralogy makes it possible to group the ore minerals according to their relative ages and to reach conclusions in regard to the nature of the processes responsible for the deposition of each group.

The solutions that deposited the great bulk of the minerals of the Tonopah ores were, it is believed, exudations from cooling, buried bodies of igneous rock or magma. In expressing this view the writers accept the opinion of Spurr for the Tonopah district and align their conclusions with the trend of opinion among American geologists in regard to the genesis of the majority of precious-metal lode deposits. The accumulation of independent evidence of such an origin would have involved structural and stratigraphic studies for which the writers had little opportunity. Such mineralization is commonly termed primary, but the writers prefer to use the term hypogene, proposed by Ransome.[2] Solutions coming from great depths within the earth and having a general upward flow will therefore be termed hypogene solutions and the work they accomplished hypogene mineralization.

After the hypogene mineralization the slow removal of the covering of volcanic rocks by erosion exposed the upper parts of many of the veins to the action of the air and of waters of surface origin. By the combined attack of these agents upon the upper parts of the hypogene ore bodies, silver and other metals were dissolved. These metals were carried downward in solution and eventually redeposited, commonly in fractures or vugs in the hypogene ores. These are the processes commonly referred to as "oxidation" and "enrichment," but they will be referred to collectively in this paper as supergene mineralization.[3]

Although supergene mineralization has played an interesting part in the genesis of certain of the Tonopah ores, it appears to have been quantitatively much less important than the hypogene mineralization. Most of the ore bodies in which it was most prominent are worked out, and the principal dependence of the mines to-day is upon ores that, the writers believe, are almost wholly of hypogene origin.

HYPOGENE MINERALIZATION.

The hypogene ores will be considered first, because they are at present the principal economic resource of the camp and are the parent type from which the supergene ores were derived.

HYPOGENE ORE TEXTURES.

The ores of Tonopah occur in veins that were formed mainly by replacement [4] of the country rocks along zones of closely spaced fracturing. Ore deposition in open spaces was an accompanying but minor process. In places, conspicuous slip planes, accompanied by much gouge, limit or traverse the ore bodies, but commonly the ore bodies are without definite walls on one or on both sides.

LARGER TEXTURAL FEATURES.

The observations here recorded on the larger textural features of the hypogene ores were mainly incidental to the microscopic studies. A complete study of the larger vein textures was not attempted.

Banding in the ores was particularly conspicuous in parts of the Murray veins and commonly presents the appearance shown in

---

[1] R. E. Liesegang (Geologische Diffusionen, Dresden, 1913) has shown that certain phases of banded structure may be produced by rhythmic precipitation during the diffusion of a solution into some medium with which it reacts chemically, or by simultaneous diffusion of two mutually interacting solutions into an inert medium. It is highly probable that some of the banding observed in ores formed by replacement has developed in this way.

[2] Ransome, F. L., Copper deposits near Superior, Ariz.: U. S. Geol. Survey Bull. 540, p. 152, 1914.

[3] Idem, p. 153. "The suggestion is offered that minerals deposited by generally downward-moving and initially temperate solutions may be termed supergene minerals."

[4] The term "replacement," as used by most economic geologists, signifies the removal of material by solution and the immediate deposition of different material in its place—either the removal of rock minerals and the deposition of ore minerals or the removal of one ore mineral and the deposition of another.

Plate I, *A*. Most of the lighter areas are a microscopic intergrowth of rhodochrosite and quartz; the darker areas are quartz carrying sulphides in a very fine state of division. Plate I, *B*, shows a similar structure in ore of identical appearance from the Belmont vein. In many places in both veins the banding is concentric around fragments of wall rock (probably silicified rhyolite), one of which, partly replaced by quartz, is shown below the

FIGURE 1.—Rudely concentric arrangement of sulphides around angular fragment of wall rock. One-half natural size. Belmont vein, 1,000-foot level. Sketched by E. S. Bastin.

center of Plate I, *B*. The demarcation of the light and dark bands is in places sharp but elsewhere vague. The bands may be a millimeter or less in width and confined to those parts of the matrix immediately bordering the fragments, or they may be fairly wide, as shown in Plate I. Scattered grains of pyrite are abundant in the rhyolite fragments but are absent from the ore surrounding them. The fragments are commonly angular, but some show evidences of gradual replacement by the fine quartz-rhodochrosite intergrowth. The arrangement of the black sulphide-rich portions around a wall-rock fragment in the Belmont vein is sketched in figure 1.

The concentric arrangement of many of the bands around angular fragments of wall rock favors their interpretation as true crustification. As banding of somewhat similar appearance may be developed in ores that have been formed solely by replacement, in the absence of any but very minute openings, it is difficult to exclude the possibility that some of the banding in the Tonopah ores may have originated during the replacement of a matrix of crushed rock in which larger fragments of wall rock were embedded. Undoubted crustification is locally observed in narrow veinlets traversing the wall rocks, as sketched in figure 2.

HYPOGENE (?) VEINLETS AND "BRECCIAS" OF PYRARGYRITE.

A number of specimens of rich ore obtained through the courtesy of Mr. John G. Kirchen show veinlets, mainly of pyrargyrite, traversing quartz and partly silicified wall rock. These specimens came from the Tonopah-Extension mine, but their exact location in the mine is uncertain. The appearance of sawed faces is shown in Plate II, *A* and *B*. As shown particularly well in Plate II, *B*, these veinlets pass from areas of only partly silicified wall rock (*a*) into areas of quartz (*b*) that apparently represent completely silicified wall rock. Upon entering the quartz the veinlets become somewhat less well defined. Thin sections across the contact between the partly silicified and completely silicified wall rock show under the microscope that the veinlets continue with little change of form from the partly altered rock into the quartz and that they were formed after the silicification that developed the quartz areas. The veinlets are planes of minor fracturing, along which pyrargyrite has been deposited along minute fractures partly filled with fragments of quartz and altered wall rock (mainly sericite and pyrite). Some pyrargyrite has replaced the sericitized fragments and some has filled small quartz-lined vugs. In places some pyrite is so intergrown with pyrargyrite as to suggest that both were deposited at the same time. Pyrargyrite is

FIGURE 2.—Crustification in small veinlet traversing wall rock. Natural size. a, Sulphides; b, quartz; c, center of veinlet. Belmont vein, 1,000-foot level. Sketched by E. S. Bastin.

not confined to the veinlets, but occurs also as small patches of irregular outline scattered through and evidently replacing the partly altered bordering rock.

Areas showing pyrargyrite veinlets of the type just described or showing minute quantities of pyrargyrite replacing altered rock are closely associated with areas as much as 2 inches across of what appears to be a breccia of pyrargyrite fragments in a gray siliceous matrix. The pyrargyrite fragments are at most 1 centimeter across. When viewed in detail under the microscope the pyrargyrite areas are seen to have irregular outlines, as shown in Plate II, *C*, and to be in the main formed by the replacement of a breccia of quartz and sericitized wall rock. The fragmental appearance is due to the fact that certain fragments in this breccia which were evidently more susceptible .to replacement than the matrix and other fragments have been almost completely replaced by the pyrargyrite.

Reliable criteria for determining whether these pyrargyrite veinlets and breccias were deposited by hypogene or by supergene solutions seem to be lacking. They may represent an early step in the extensive replacement of wall rock by an association of quartz and pyrargyrite of the type shown in Plate III, *A*, which may be very uniform in character through several cubic inches and is regarded by the writers as probably hypogene. Furthermore, most deposits of pyrargyite and other silver minerals in late fractures in the Tonopah ores can be cleanly stripped from the walls of such fractures, and the walls appear to have undergone no replacement. Such deposits are interpreted as supergene and are quite different from the pyrargyrite deposits described in the present section. The writers are inclined to believe, therefore, that these unusual pyrargyrite veinlets and breccia-like deposits are hypogene, though recognizing that the evidence is not conclusive.

MINUTE TEXTURAL FEATURES.

The microscopic study of the ores reveals minute textural relations, not otherwise recognizable, that are of great importance in interpreting their genesis. It shows repeatedly and with especial clearness that the primary or hypogene mineralization was not a simple, brief event but was a process of considerable duration accomplished by solutions whose composition was not at all times the same. Ore minerals deposited early in the hypogene mineralization were partly or wholly dissolved at a later stage in the process, and other ore minerals more stable in the presence of the changed solutions were deposited in their place.

Some of the general features of the hypogene mineralization and the principles which they exemplify may first be considered. The Tonopah ores are in the main formed by the replacement of igneous rocks; these rocks are aggregates of many mineral species, some of which are more susceptible to replacement, under a given set of conditions, than others. The effects of the first of the hypogene solutions, the advance wave of mineralization, were the replacement of certain of the wall-rock minerals by other nonmetallic minerals and by pyrite, alterations commonly included under the term "hydrothermal alteration." These changes are now shown in the wall rock bordering the veins and probably were accomplished in the main vein zones themselves, where they have since been obliterated by further changes. They have been carefully studied by Spurr.[1] In the Mizpah trachyte, the predominant wall rock of many of the veins, the first alterations commonly consisted in the selective replacement of certain minerals by calcite, siderite, chlorite, sericite, and pyrite; later there was progressive destruction of chlorite and calcite and to some extent of pyrite, accompanied by an increase in the amount of siderite and sericite and the development of quartz and adularia; finally the pyrite was completely redissolved and the rock converted into an aggregate of quartz and silica. These changes, while broadly a part of the mineralization process, preceded the introduction of the valuable metals, for, according to Spurr,[2] the pyritized wall rocks carry only negligible amounts of gold and silver. The present paper is concerned with the succeeding changes, which involved the further replacement of the altered rocks and the development of the typical ore minerals. Mainly for convenience in description and without attributing any particular significance to such a classification, the writers have divided the hypogene ore minerals into two classes—alpha hypogene and beta hypogene.

The alpha hypogene minerals are those which are formed by the direct hypogene replacement of wall-rock minerals or their common hydrothermal alteration products, such as sericite and quartz, or which are not demon-

[1] Spurr, J. E., Geology of the Tonopah mining district, Nev.: U. S. Geol. Survey Prof. Paper 42, pp. 207–252, 1905.
[2] Idem, p. 207.

strably otherwise. The beta hypogene minerals are formed by the replacement of earlier hypogene metallic ore minerals. Replacement of the host rock may go on until it is practically complete, after which further replacement must involve the destruction of earlier ore minerals. The beta hypogene minerals are obviously younger than those particular alpha hypogene minerals which they replace, but they are not necessarily younger than all alpha hypogene minerals. On the contrary, there is much evidence that direct replacement of the host rock by ore minerals at one place was coincident with replacement of ore minerals by other ore minerals at another place. For this reason the terms "alpha" and "beta" are preferred to terms that carry a time significance, such as "early" and "late."

ALPHA HYPOGENE MINERALS.

GENERAL RELATIONS.

The individual grains of the minerals that are believed to be alpha hypogene interlock irregularly in a fashion indicative of mutual interference during growth and produce ore textures somewhat comparable to the granular textures of certain igneous rocks. Typical examples of these textures are shown in figures 3 to 5 and in Plates V, A, and VI, B. Such textures are indicative of general contemporaneity of the minerals present, although certain minor age differences probably exist, such as are shown by the order of crystallization usually recognizable among the components of a massive igneous rock.

The minerals named below appear to have been formed by alpha hypogene crystallization in nearly all the veins studied—Belmont, Favorite, Shaft, MacDonald, North Star footwall vein, Last Chance, West End, Extension combined vein on 770-foot level, Egyptian, Extension lower contact vein, McNamara, and Murray:

Sphalerite.	Argyrodite (?).
Galena.	Polybasite.
Chalcopyrite.	Argentite.
Pyrite.	Electrum.
Arsenopyrite (noted only in the Last Chance vein).	Quartz.
	Carbonates.
Pyrargyrite.	

Carbonates were noted in all these veins, but their composition commonly differs from point to point even within a single vein, as explained in the paragraph on carbonates.

SPHALERITE.

Sphalerite, which is one of the commonest sulphides of the ores, was noted only as an alpha hypogene mineral. Crystal faces on

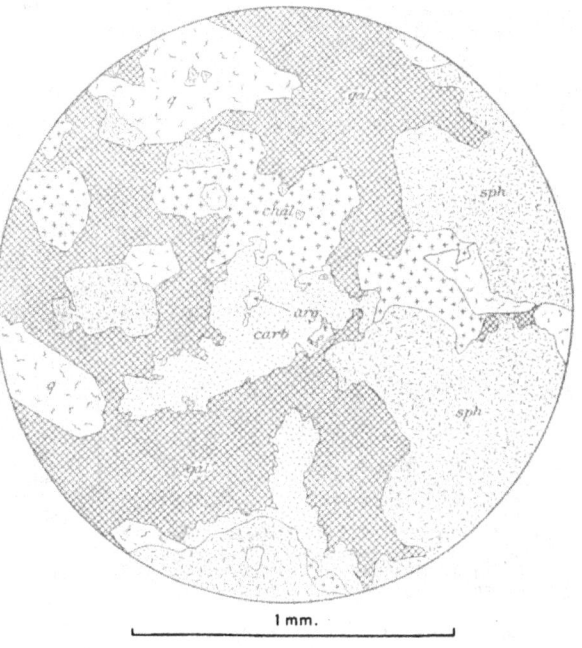

FIGURE 3.—Typical association of alpha hypogene ore minerals. A few small areas of argentite (arg) may be later replacement deposits, but the other minerals are believed to be essentially contemporaneous. q, Quartz; gal, galena; chal, chalcopyrite; sph, sphalerite; carb, carbonate; arg, argentite. Camera lucida drawing from polished surface of ore from Favorite vein, Tonopah Belmont mine. Depth uncertain.

sphalerite were noted only in a few places where the mineral was in contact with galena or chalcopyrite. Commonly it forms rounded areas, which are surrounded (see fig. 4) by one or more of the minerals galena, argentite, or chalcopyrite. Such rounded outlines might suggest that the sphalerite has been partly replaced by the minerals that surround it, but there are no other indications that sphalerite has been replaced to more than a very minor extent during the entire process of mineralization. Locally, as shown in Plate III, B, the sphalerite has very irregular outlines. This plate also illustrates a less common mode of occurrence of sphalerite, as small, irregular veinlets that appear to have been formed by replacement of the host rock along minute fractures. The boundaries between sphalerite and quartz are generally very irregular (see fig. 5) and suggest complete contemporaneity of deposition. In a few places sphalerite is traversed by minute replacement veinlets of chalcopyrite and polybasite, as is shown in Plate VIII, *A*, and figure 7.

In general the maximum deposition of sphalerite and quartz appears to have occurred early in the hypogene mineralization, and the maximum deposition of galena, chalcopyrite, pyrargyrite, and argentite was slightly later, as is shown by their matrical position. The two periods overlapped, however, and locally sphalerite and quartz are later than the other

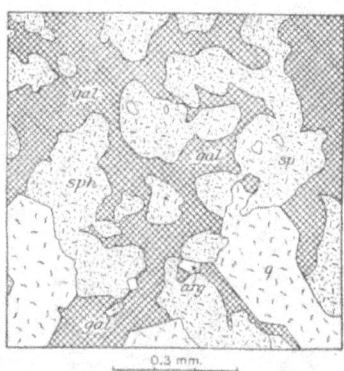

FIGURE 4.—Sphalerite (*sph*) with rounded outlines surrounded by galena (*gal*). A few small areas of argentite (*arg*). *q*, Quartz. Camera lucida drawing from polished surface of ore from Favorite vein, Tonopah Belmont mine. Depth uncertain.

minerals. In most of the ores sphalerite is not identifiable without the aid of the microscope.

GALENA.

Galena, which is also one of the commonest sulphides of the ores, was noted only as an alpha hypogene mineral. It almost invariably conforms to the crystal outlines of bordering quartz areas, as shown in figure 3, and in many places it forms the matrix of areas of sphalerite of rounded outline, as shown in figure 4. Its contacts with alpha hypogene chalcopyrite and argentite are commonly irregular or smoothly curving. Galena has been replaced by beta hypogene minerals to a greater extent than any other sulphide.

CHALCOPYRITE.

Alpha hypogene chalcopyrite is nearly as abundant as sphalerite or galena. It also occurs as a beta hypogene and probably as a supergene mineral, and it is not always possible to determine to which class a given occurrence belongs. Chalcopyrite in relations like those shown in figure 3 is interpreted as alpha hypogene. The mineral commonly conforms to the crystal outlines of quartz and locally embeds sphalerite of rounded or crystal outline. Its boundaries against galena and alpha hypogene argentite are commonly irregular or rounding. Alpha hypogene chalcopyrite has locally been replaced by beta hypogene minerals and by supergene minerals.

PYRITE.

Pyrite is in general less abundant in the ores than chalcopyrite, but tends to form larger grains. It was the first sulphide to form in the process of mineralization by replacement, and in slightly mineralized rock it is the only one present. (See Pl. XV, *B*.) Its deposition appears to have continued throughout the hypogene mineralization, for in many places it conforms to the crystalline outlines of quartz and has irregular boundaries next to chalcopyrite.

Pyrite appears to have formed locally as a late deposit in open fractures but was nowhere noted as having replaced other metallic ore minerals.

ARSENOPYRITE.

A mineral that is nearly white in the polished sections and only slightly softer than pyrite, though it takes a smoother polish, is probably arsenopyrite. The outlines of the crystal indicate orthorhombic crystallization. Stellar trillings, such as are common in arsenopyrite, were noted. Though the mineral is fairly abundant in an ore specimen from a depth of about 500 feet in the Last Chance vein, it was not noted in any other specimens. It is irregularly intercrystallized with sphalerite, galena, and quartz, and except for its presence the ore in which it is found presents no unusual features.

ARSENICAL PYRARGYRITE.

Arsenical pyrargyrite, though rare compared with the sulphides of the base metals, is present in all the veins studied and is locally very abundant. Spurr [1] appears to have regarded it as invariably supergene. Some of the pyrargyrite almost certainly is of supergene deposition,

[1] Spurr, J. E., Geology of the Tonopah mining district, Nev.: U. S. Geol. Survey Prof. Paper 42, pp. 94–95. 1905.

as will be shown later, but most of it appears to be hypogene.

Where particularly abundant its chief associates are quartz, polybasite, and probably argyrodite, but in places, as shown in figure 5, it is irregularly intercrystallized with the common sulphides, sphalerite, pyrite, chalcopyrite, etc. Veinlets of pyrargyrite that may be alpha hypogene are described on page 11 and illustrated in Plate II. More irregular intergrowths. with quartz of the type shown in Plate III, A, were noted in ores from the Last Chance and Fraction veins and the Tonopah Extension mine. A texture of this sort might be explained through the deposition of pyrargyrite in the open spaces of a porous mass of finely crystalline quartz, but it is more probably the result of simultaneous replacement of a rock by quartz and pyrargyrite, for analogous textural relations exist elsewhere between quartz and the commoner sulphides or mixtures of the commoner sulphides with pyrargyrite, as shown in figure 5. In ores like those shown in Plate III, A, and figure 5 the quartz possessed greater power of assuming crystal outlines than the other ore minerals. The microscopic intergrowths of quartz and pyrargyrite may constitute several cubic inches of material which, viewed with the unaided eye, appears to be pure pyrargyrite.

A few small crystals of pyrargyrite occur in vugs, and the angles of some of these have been measured by Eakle.[1]

A large number of samples of pyrargyrite heated in closed tubes before the blowpipe carried arsenic, and in some samples arsenic appeared to be as abundant as antimony.

Pyrargyrite was not noted as replacing other ore minerals (beta hypogene).

Louis G. Ravicz[2] in a recent paper expresses the view that the sulphosalts of silver, such as pyrargyrite and polybasite, can not be formed by precipitation from solutions containing alkaline sulphides through decrease of pressure or temperature alone, because under ordinary

[1] Eakle, A. S., The minerals of Tonopah, Nev.: California Univ. Dept. Geology Bull., vol. 7, p. 9, 1912.
[2] Experiments in the enrichment of silver ores: Econ. Geology, vol. 10, p. 375, 1915.

pressures they are unstable in hot solutions of alkaline sulphides. He suggests that whenever these minerals are found in apparently primary occurrences their precipitation has resulted from the mingling of ascending alkaline sulphide solutions with descending acid solutions. It appears to the writers inherently improbable that oxidation could be furnishing any significant amount of acid water at a time when hot mineralizing solutions were still ascending. Unless it is assumed that mineral deposits from hypogene solutions are nowhere being formed at the present day it must be supposed that certain hot springs (not all or even the majority) are the surface expression

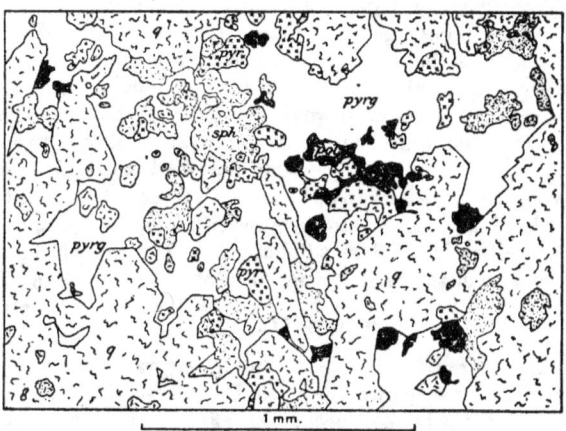

FIGURE 5.—Irregular intergrowth of pyrargyrite (pyrg) with quartz (q), sphalerite (sph), pyrite (pyr), and polybasite (pol). Camera lucida drawing from polished surface of ore from Fraction vein, Jim Butler mine, between 600 and 700 foot levels.

of mineralization in progress below them. Mineral deposits being formed at or near the surface in regions of hot-spring activity are characteristically rich in gangue and poor in metallic minerals; moreover, the metallic minerals are confined to certain species, such as stibnite and cinnabar, and rarely, so far as the writers are aware, include the common sulphides galena, sphalerite, and chalcopyrite, which are the predominant sulphides at Tonopah and are locally intergrown contemporaneously with the sulpho-salts of silver. The composition of ores known to be formed at the surface appears therefore to be not the most favorable for the development through oxidation of large amounts of acid solutions, even if the upflow of ascending hot waters did not inhibit extensive oxidation.

Furthermore, the mineralogy of the Tonopah ores is not that of the shallowest types of ore deposits. Mining has shown that the hypogene ores range through a vertical interval of 1,500 feet without notable change in mineral composition, and it seems probable that they were originally deposited at several times this depth, at horizons where effective mingling with acid supergene solutions probably could not occur. If Ransome[1] has correctly interpreted the formation of alunite at the neighboring camp of Goldfield as the result of a mingling of descending acid solutions with the ascending solutions, the absence of alunite at Tonopah may be taken as additional evidence that such intermingling did not take place there.

Although the writers are not inclined to regard the mingling of supergene with hypogene solutions as the cause of the precipitation of the hypogene sulphosalts of silver at Tonopah, these minerals are in the main the latest of the hypogene minerals, and it is quite probable that at the time they were deposited the hypogene solutions had become somewhat depleted in alkalies, through sericitization of the wall rocks, and in sulphur, through precipitation of the early hypogene sulphides.

ARGYRODITE (?).

In intergrowths of quartz and pyrargyrite, such as those shown in Plate III, A, small amounts of polybasite occur, and also a metallic mineral that could not be certainly identified. This mineral, which is irregularly intergrown with the pyrargyrite and polybasite and is plainly contemporaneous with them, is recognizable only upon microscopic study of polished specimens. On freshly polished surfaces it is distinguishable from the pyrargyrite, though not easily, by its slightly pinkish tint. It is brittle and near pyrargyrite in hardness, but has a black streak about like that of polybasite. It is not tarnished by silver nitrate or mercuric chloride solutions that tarnish pyrargyrite and polybasite brown. The mineral was found in too small amounts to be isolated for analysis, but its physical properties appeared to agree with those of the mineral argyrodite (Ag_8GeS_6 or Ag_6GeS_5), in which the element germanium was first discovered. Chase Palmer, of this Survey, has made careful qualitative tests for germanium in a sample of the mixture which carried the unknown mineral and definitely determined that this element was present. It is therefore probable that the mineral argyrodite is present in these ores, although the possibility that the germanium may be in some other mineral combination can not be excluded.

In the two localities in which argyrodite has been identified, it is also associated with other silver minerals. Near Freiberg, Saxony, it is associated with pyrargyrite, polybasite, stephanite, and common sulphides of the base metals.[2] In Bolivia it is associated with polybasite and stephanite.

POLYBASITE.

The mode of occurrence of polybasite is very similar to that of pyrargyrite, with which it is commonly associated. It occurs not only as an alpha hypogene but also as a beta hypogene and a supergene deposit. In its alpha hypogene occurrence it may be irregularly intergrown with pyrargyrite and the commoner sulphides, as shown in figure 5, or may be a component of veinlets of the type illustrated in Plate II, A and B. In a few deposits of the type shown in Plate IV, A, the polybasite, probably hypogene, is intimately intercrystallized with quartz and pyrite.

In one specimen from the Tonopah Extension mine polybasite inclosed by pyrargyrite showed the blade-like forms illustrated in Plate IV, B. These blades are presumably cross sections of polybasite crystals.

STEPHANITE (?).

The presence of stephanite in the Tonopah ores has not been proved. Numerous tests were made on crystals that from general appearance might be either stephanite or polybasite, to determine whether they contained copper. Upon dissolving in nitric acid and neutralizing with ammonia the blue color of the copper-ammonium salt was obtained in every test. This was not due to admixed chalcopyrite, for no precipitate of ferric hydroxide formed. All these crystals were therefore identified as polybasite. Spurr[3] and Eakle[4] also

---

[1] Ransome, F. L., The geology and ore deposits of Goldfield, Nev.: U. S. Geol. Survey Prof. Paper 66, pp. 193-195, 1909.

[2] See Hintze, Carl, Handbuch der Mineralogie, Band 1, Abt. 1, pp. 1193-1197, 1904, for full references to the literature.

[3] Spurr, J. E., Geology of the Tonopah mining district, Nev.: U. S. Geol. Survey Prof. Paper 42, p. 95, 1905.

[4] Eakle, A. S., op. cit., p. 9.

were unable to identify any of the "brittle silver" as stephanite.

ELECTRUM.

A pale-yellow mineral that is sectile and is not tarnished by HCl, HNO$_3$, or (NH$_4$)$_2$S is recognizable under the microscope in a large number of the ores. It occurs in grains so fine that its mechanical separation from the other ore minerals is extremely difficult. Ore from the West End vein between the 400 and 500 foot levels that was known from microscopic study to carry this mineral was very finely crushed and carefully panned. After the heavy pannings were treated with hot dilute nitric acid to dissolve the sulphides, numerous heavy minute black particles, visible under a hand lens, remained. Upon further treatment with hot concentrated nitric acid these particles acquired the yellow color characteristic of gold. Similar blackening of the mineral upon treatment with dilute acid was also noted by Hillebrand. Several specimens shown by microscopic study to be rich in the light-yellow mineral assayed exceptionally high in gold. This mineral is believed to be electrum, possibly carrying some selenium, as discussed below. This conclusion accords with the studies of Hillebrand on ores collected by Spurr, which showed that at least some of the gold in the ores was heavily alloyed with silver.

A number of assays of Tonopah ores exceptionally rich in gold are given below. How much of the silver in these ores was alloyed with gold is unknown, for sulphides and sulpho-salts of silver were present in all the samples.

Assays of some Tonopah ores.

[Ounces per ton.]

	Gold.	Silver.
1. North Star mine footwall vein, 1,250-foot level; grab sample from 250 tons.	47.96	19.35
2. Same locality as No. 1; picked piece of rich ore.	367.22	185.15
3. Same locality as No. 1; two pieces.	79.60	36.20
4. Same vein as No. 1, stope above 1,130-foot level.	10.72	681.90
5. West End vein in stope 522 between 500 and 600 foot levels; small piece of rich ore.	23.50	1,483.00

Assays 1 to 4 above were communicated by Mr. A. E. Lowe, superintendent at the North Star mine. Sample 2 is said to have shown no free gold, even on panning; presumably the gold was in very minute grains, which were not freed from the sulphides in the crushing preparatory to panning. Microscopic examination by the writers of ore said to be from the lot represented in assay 4 revealed the presence of the light-yellow sectile mineral in unusual abundance and in unusually large grains (as much as 0.36 millimeter in diameter). A part of the silver in this ore occurs in pyrargyrite, polybasite, and argentite, all of which were identified microscopically. Assay 5 was made by the Bureau of Mines on a small sample collected by the writers and shown by microscopic study to be unusually rich in the light-yellow mineral. Pyrargyrite, argentite, and polybasite were also recognized microscopically in this sample.

In most of the ores the electrum grains conform to the characteristic rounded boundaries of sphalerite and to crystal boundaries of quartz, sharing the matrical position of galena, chalcopyrite, and argentite. (See Pl. V, *A*.) In places it develops crystal faces. (See Pl. V, *B*.) The argentite shown in Plate V, *B*, is believed to have been formed, at least in part, by the replacement of galena, for reasons discussed in the section on argentite (p. 18). It is possible that the electrum in this section also replaces galena, but more probably it crystallized contemporaneously with the galena whose place is now occupied by argentite. Exceptionally the electrum is intercrystallized irregularly with sphalerite or bordered by it on all sides in the plane of the section. In a number of ores it was observed as a minor component of the carbonate-sulphide aggregates. (See p. 18 and Pls. VI, *B*, and VII, *A*.)

SELENIUM.

Small amounts of selenium have long been known to be present in the ores of Tonopah and in the derived concentrates and bullion. The largest amount recorded is 2.56 per cent, in an analysis by W. F. Hillebrand of pannings from heavy sulphide ore (analysis 2, p. 46). The mineral combination in which it occurs is not yet definitely determined, but there is considerable ground for believing that it is associated with the gold and silver in electrum. Mr. Palmer definitely determined the presence of selenium in pannings consisting mainly of electrum, from ore obtained in the West End mine; unfortunately the quantity of material available was very small. It is possible that selenium produced the blackening in the electrum upon treatment with dilute nitric acid, noted above.

ARGENTITE.

The extent to which argentite occurs as an alpha hypogene mineral is somewhat uncertain. In a few places it is associated with sphalerite in the irregular fashion shown in Plate III, B, and in such places it appears to be alpha hypogene. Small areas of argentite bordered on all sides, in the plane of the section, by sphalerite, as shown in figure 4, also appear to be alpha hypogene. Argentite is also an abundant component of the fine carbonate-sulphide intergrowths described below, and in some of these it probably directly replaces the host rock and is therefore alpha hypogene.

One of the commonest modes of occurrence of argentite is shown in figure 10 and Plate V, B. In the specimens illustrated its position is interstitial with respect to quartz with crystal outlines and entirely similar to that commonly assumed by galena. (See fig. 4.) As a rule it is possible to prove that this argentite is at least in part formed by the replacement of galena, but some of it may directly replace the host rock.

Argentite is also abundant as a supergene mineral.

QUARTZ.

Quartz is the most abundant gangue mineral of the ores and is universally present. Most of it appears to have been formed by direct replacement of the host rock. In a few places it appears to have been replaced by the carbonate-sulphide aggregates described below, but replacement of quartz has occurred only to a very minor extent in this district.

Thin coatings of quartz, locally found on the walls of certain open fractures, are believed to be in part supergene and in part very late hypogene deposits.

CARBONATES.

The pale-pink material so abundant in many of the Tonopah ores has commonly been ermed rhodonite, on account of its apparent hardness, the fact that it effervesced with acid being attributed to admixed carbonates. Microscopic study shows, however, that it is a very fine intergrowth of quartz and a carbonate, and that no rhodonite is present.

An analysis by W. F. Hillebrand [1] of sulphide ore from depths of 460 to 512 feet in the Montana Tonopah mine showed that four elements were combined as carbonates in the following proportions: Calcium 2.7, magnesium 0.9, iron 1.14, and manganese 1.32. Qualitative tests by the writers on pink carbonate from the 906-foot level on the Murray vein showed the presence of calcium, magnesium, manganese, and a very small amount of iron. Carbonate of similar appearance from the 1,260-foot level on the same vein carried a much larger proportion of iron. The fine state of division of the carbonate in the ores makes it difficult to determine whether several hypogene carbonates are present in varying proportions or a single carbonate of varying composition.

Much of the carbonate is present in the fine intergrowths with sulphides, which are described in the next section, but radiating intergrowths with quartz of the type shown in Plate VI, A, are common, especially in banded ores like those shown in Plate I.

In some veins, such as the Murray, intergrowths of quartz and pink carbonate are very abundant and give a pink cast to much of the ore; in other veins, such as the West End and Favorite, hypogene carbonates are usually visible only with the aid of the microscope.

Carbonates also occur as supergene deposits.

CARBONATE-SULPHIDE AGGREGATES.

Microscopic aggregates of light-colored carbonates with one or more of the minerals chalcopyrite, galena, argentite, and polybasite are common in many of the ores and are shown in Plate VI, B. Usually only one of the silver minerals is present. In general, all the minerals of these aggregates conform to the crystal outlines of quartz and to the rounded outlines of sphalerite. The fine aggregates are, as a rule, sharply demarked from the bordering coarser material, but exceptionally there are gradual transitions from fine aggregates in which carbonates predominate to coarser intergrowths in which sulphides predominate.

[1] Spurr, J. E., Geology of the Tonopah mining district, Nev.: U. S. Geol. Survey Prof. Paper 42, p. 89, 1905.

Such a transition from areas consisting mainly of carbonate with small amounts of galena and argentite to areas consisting of mainly galena and argentite with scattered patches of carbonate is shown in Plate VII, *A*. Transitions into areas consisting mainly of chalcopyrite were noted in other specimens. These transitions, as well as the matrical position of the carbonate-sulphide aggregates, indicate that they were deposited for the most part contemporaneously with the larger areas of galena and chalcopyrite. As already stated, a few grains of electrum occur in some of these aggregates.

WOLFRAMITE.

The tungstate of iron and manganese has been noted at a number of localities in the mines at Tonopah. The writers are not completely informed as to all its occurrences, but it was noted by them in the Belmont vein and in a small vein in the Valley View mine.

In the Valley View mine, on the 300-foot level, quartz and wolframite form a small vein as much as 4 inches wide, lying well within the footwall of the Valley View vein. Its relations to the silver veins are not shown.

In the Belmont vein as exposed on the 1,000-foot level of the Tonopah Belmont mine, in south crosscut 1044, wolframite is very abundant and its relation to the silver ore is well exhibited. The silver ore at this locality consists mainly of quartz and an intergrowth of quartz and ferruginous rhodochrosite of the sort described on page 18, through which aggregates of very finely divided sulphides are distributed. Fractures traversing this ore are partly or completely filled with crystalline quartz, with which wolframite is so intimately intergrown as to leave no doubt that the quartz and wolframite are strictly contemporaneous. Wolframite is also intercrystallized with quartz that lines vugs in quartzose parts of the ore.

So far as the writers are aware wolframite has nowhere been established as a mineral of supergene deposition. The wolframite at Tonopah is believed to have been deposited from ascending silica bearing waters after the deposition of most of the silver ores and after these ores had undergone some fracturing.

BETA HYPOGENE MINERALS.

GENERAL FEATURES.

Certain minerals and mineral aggregates of the Tonopah ores are formed by the replacement of earlier ore minerals rather than by direct replacement of the wall rock or of the products of its hydrothermal alteration. The criteria by which those deposits that replace earlier ore minerals may be recognized in these ores are here briefly summarized. Some of the deposits are believed to be hypogene and others supergene, and the criteria for distinguishing between them are discussed on pages 24 and 42.

The textural relation of the replacing minerals of greatest diagnostic value is their

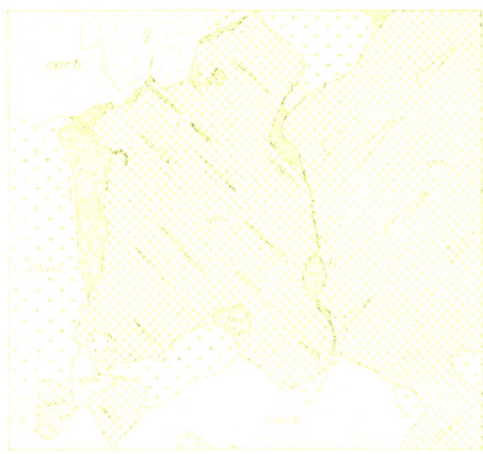

FIGURE 6.—Replacement of galena (*gal*) by polybasite (*pol*) along its contacts with other minerals (as at *A* and *A'*), along contacts between two differently oriented galena crystals (as at *B*), and along galena cleavages (as at *C*). Rims of an earlier replacement product like those shown in figure 14 are invariably present between polybasite and galena but are too narrow to be shown on this scale. *q*, Quartz; *carb*, carbonate; *chal*, chalcopyrite; *sph*, sphalerite. Camera lucida drawing from polished surface of ore from Favorite vein, Tonopah Belmont mine. Depth uncertain.

localization at places that were obviously channels of easy access for mineralizing solutions. Situations particularly favorable for the replacement of hypogene ore minerals by other ore minerals are the following:

1. Contacts between hypogene minerals of different species, at least one of which was replaceable under the then existing conditions. In such situations the replacing minerals

commonly form narrow replacement bands, as shown at A and A' in figure 6 and in Plate VIII, B.

2. Contacts between two crystals of the same replaceable mineral. The replacing mineral along such contacts commonly takes the form of narrow bands expanding here and there into lenticular areas, as is well shown at B in figure 6 and A in Plate VII, B.

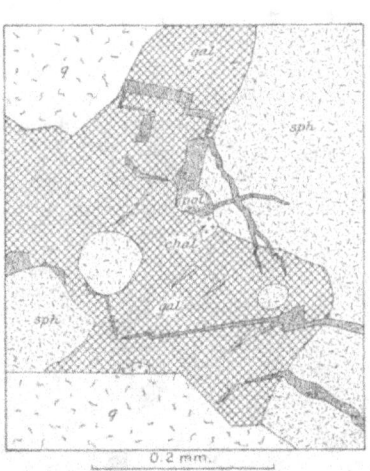

FIGURE 7.—Replacement of galena (gal) and sphalerite (sph) by polybasite (pol) and chalcopyrite (chal). In sphalerite the replacement veinlets are commonly irregular, but in galena many of them follow cleavage directions. Locally the replacement has followed the contacts between galena and other minerals. q, Quartz. Camera lucida drawing from polished surface of ore from Belmont vein, 1,200-foot level of Tonopah Belmont mine.

3. Narrow wedges of replaceable mineral between converging contacts with other minerals or between intersecting fractures. These are particularly vulnerable to replacement, as is well illustrated at A in figure 9.

4. Cleavage planes in replaceable minerals or obscure fractures following cleavages. Replacement of galena along cleavages is shown at C in figure 6 and in figure 7.

5. Minute irregular fractures. Replacement along irregular fractures in sphalerite is shown in figure 7 and in Plate VIII, A.

The replacement of alpha hypogene minerals appears not to have been confined to situations of obvious vulnerability but in many places seems to have proceeded irregularly through the ore. Some deposits so formed represent merely advanced stages of a process which began by replacement along mineral contacts or fractures, but in reference to others it seems necessary to conclude that the ore was so thor-oughly "soaked" with the replacing solutions that replacement proceeded with great facility, without much regard to the earlier textural features of the ore. Instances of replacement of this sort will be cited in the detailed descriptions. In some places the resulting textures can be identified as due to replacement only by local gradations into undoubted replacement deposits along mineral contacts and fractures or by close textural and mineralogic resemblance to such deposits.

The minerals believed to have been formed by hypogene replacement of earlier hypogene ore minerals are polybasite, argentite, an undetermined mineral that is probably a lead-silver sulphide, chalcopyrite, electrum, and a fine-grained carbonate-sulphide aggregate.

POLYBASITE.

Polybasite is very common in the ores of Tonopah, replacing galena and, very rarely, sphalerite. The situations at which replacement may occur are well shown in figures 6, 7, and 8. Along the contacts between galena

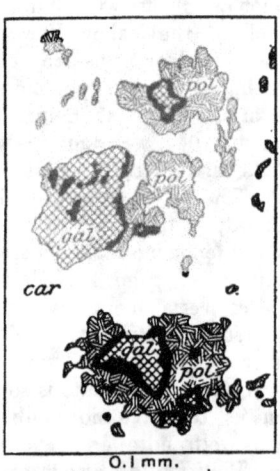

FIGURE 8.—Small areas of galena (gal) in the fine carbonate-sulphide aggregates partly replaced by polybasite (pol). The first product of the replacement, possibly a lead-bearing silver mineral, forms narrow rims (black in the drawing) between the galena and the polybasite. car, Carbonate. Camera lucida drawing from polished surface of ore from Favorite vein, Tonopah Belmont mine. Depth uncertain.

crystals the replacing mineral commonly appears in the polished sections as minute threads (as at A, Plate VII, B), which enlarge to lenses at intervals (as at B, fig. 6).

Very rarely polybasite occurs alone or with chalcopyrite as minute replacement veinlets in sphalerite, as is shown in Plate VIII, *A*. These veinlets are commonly irregular in the sphalerite, but upon entering neighboring galena grains they conform to the cleavage planes of the galena. These features are shown in figure 7, a drawing, on a larger scale, from a portion of the specimen shown in Plate VIII, *A*.

In places even minute areas of galena in the carbonate-sulphide aggregates (see p. 18) have been partly replaced by polybasite, as is shown in figure 8.

Between galena and the polybasite which has replaced it, but not between sphalerite and polybasite, there is invariably a narrow band of a third mineral which is evidently a temporary transition product in the replacing process. This appears to be the same mineral that is developed between galena and argentite. It is considered more fully below.

ARGENTITE.

Although argentite occurs as a supergene deposit on the walls of open fractures, it has been formed principally by the replacement of galena. For reasons cited on page 24 most such replacement deposits are believed to be hypogene, although in general appearance they are similar to others that are probably supergene. Typical replacement of galena by argentite is shown in figures 9 and 10 and in Plates VIII, *B*, and IX, *B*. Relations like those shown in figure 10, in which argentite embeds quartz and sphalerite, are particularly common, and are in many places demonstrably the result of replacement of galena by argentite. Areas in which argentite without galena embeds quartz and sphalerite may be traced continuously into others like that shown in figure 10, in which both sulphides are present, the argentite replacing the galena, in part peripherally, with the invariable development between argentite and galena of narrow bands of an intermediate replacement product. In a few places argentite, alone or with chalcopyrite, replaces polybasite (probably alpha hypogene), as is shown in figure 11.

UNDETERMINED MINERAL, PROBABLY A LEAD-BEARING SILVER SULPHIDE.

Between galena and argentite or polybasite that has replaced it there is invariably a band of a replacement mineral representing an intermediate step in the alteration. These transition bands, which are of varying width, are particularly well shown in figure 12 but also appear in figures 8 and 14 and Plate VIII, *B*. Other illustrations of such replacement deposits do not show these bands because they are too narrow to appear on the scale used. They are not formed between galena and chalcopyrite that replaces it or at

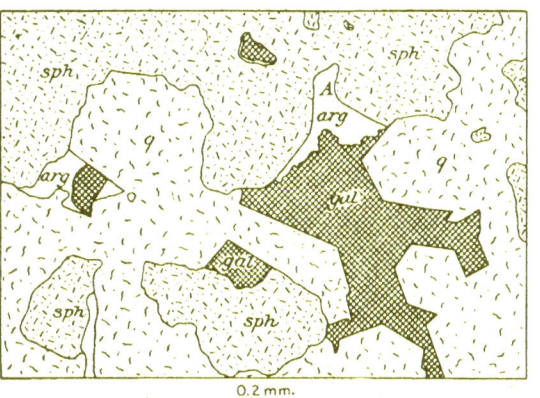

FIGURE 9.—Replacement of galena (*gal*) by argentite (*arg*) at *A*. *q*, Quartz; *sph*, sphalerite. Camera lucida drawing from polished surface of ore from Belmont vein, 1,200-foot level of Tonopah Belmont mine.

the contacts of silver minerals with any other sulphide save galena.

Here and there the replacement has proceeded no further than the formation of the undetermined mineral, as is shown in figure 13. Most of the replacement borders shown in Plate VIII, *B*, consist of the undetermined mineral, but in a few places, as at *A* and *A'*, argentite has formed as the final product.

The transition mineral is indistinguishable from galena on freshly polished surfaces, but upon treatment with hydrogen peroxide the galena is tarnished brown while the transition mineral is unaffected. Its hardness, tested by drawing a needle across the contact of galena and the transition mineral under a low power of the microscope, appears to be about the same as that of galena. The triangular depressions developed in the galena in the grinding process continue into the unidentified mineral, as shown in figures 12 and 14. These depressions are due to the tearing out of small inverted pyramids of the minerals between three intersecting cleavage planes, and their continuity from galena into the other mineral shows that the latter possesses a cleavage identical with and continuous with that of

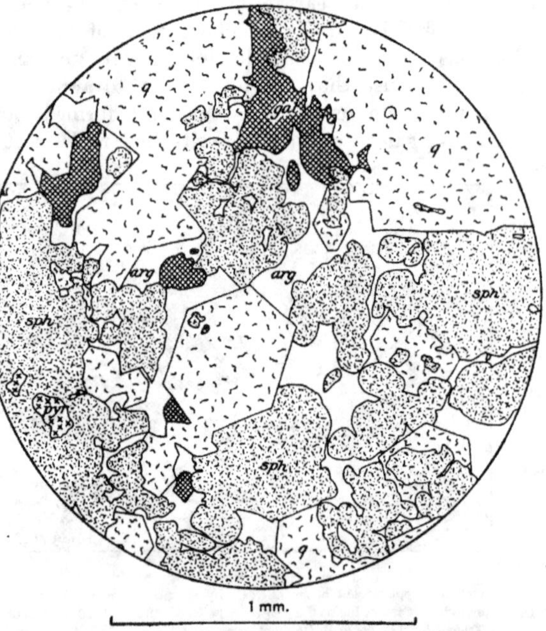

FIGURE 10.—Areas of sphalerite (sph) with rounded outlines and of quartz (q) with crystal outlines in a matrix consisting mainly of argentite (arg). These relations appear to have resulted from the replacement of galena (gal, alpha hypogene) by argentite (probably beta hypogene) in ore whose original appearance was like that shown in figure 4. Numerous remnants of galena remain, and between these and the argentite narrow rims of an intermediate replacement product (too narrow to show on this scale) are invariably present. pyr, Pyrite. Camera lucida drawing from polished surface of ore from Favorite vein, Tonopah Belmont mine. Depth uncertain.

galena. The undetermined mineral is therefore probably, like galena, isometric.

The mineral is readily distinguished from argentite by its lighter color and the fact that it is brittle rather than sectile. It is present in too small amounts and is too intimately associated with other minerals to be separated for analysis. From the properties already described it is believed to be an isometric mineral with perfect cubical cleavage, in hardness, color, and luster very similar to galena. From its position as a transition

product between galena and argentite it may be suspected of being a double sulphide of lead and silver or a solid solution of galena and argentite with physical properties like those of galena.

Chase Palmer, of this Survey, has studied experimentally the action of dilute solutions of silver salts on galena from Joplin, Mo., for the purpose of throwing some light on the character of the transition mineral observed in the Tonopah ores. In his first experiment galena, carefully picked to obtain as pure material as possible, was finely pulverized and treated in a flask with a solution of silver sulphate (about N/40) to which had been added a slightly acid solution of sodium acetate to keep in solution any lead that might be dissolved. The flask was heated in a water bath for two weeks, a stream of purified carbon dioxide passing continuously through it to prevent oxidation by air. At the end of the experiment no metallic silver could be detected by microscopic examination of the solid residue, but it was found that small amounts of lead had been dissolved and some silver removed from the solution.

For a more detailed study of the reaction between galena and a soluble silver salt, Mr. Palmer chose the organic salt silver-benzol sulphonate, which, though resembling the sulphate in containing the SO_3 radicle, is much more soluble and yields a soluble lead salt by reaction with galena. The results are stated by Mr. Palmer [1] as follows:

The galena, obtained from the Joplin district, Mo., was carefully picked. It was reduced to grains that were collected on a sieve with 40 meshes to the inch. Particles of galena were selected from these small fragments. The mineral thus chosen was ground to pass through a 200-mesh sieve.

Analysis of the powdered galena.

		Required for PbS.	
Lead.........	86.56	Lead.........	86.60
Sulphur.....	13.39	Sulphur.....	13.40
Iron........	Trace.		
	99.95		100.00

[1] Manuscript report, May 16, 1916.

For the experiment, 0.5145 gram of the powdered galena was digested in a solution of silver-benzol sulphonate containing 2.075 grams, slightly acidified with benzol sulphonic acid. The volume of the solution was about 250 cubic centimeters. The mineral was digested in this solution five days on the steam bath. The solution was then filtered. No trace of sulphate was found in the filtered solution, an indication that the sulphur of the galena had not been oxidized.

After the removal of the excess of silver from the solution, it was found that 0.11379 gram of lead had been dissolved. This represents 22.18 per cent of the galena used in the experiment. The residue was found to contain 13.38 per cent of sulphur—that is, all the sulphur present in the galena used—and from the residue was obtained 0.11242 gram of silver, representing 21.85 per cent of the weight of the galena.

$$21.85 \div 107.9 = 0.2025 \text{ silver.}$$
$$22.18 \div 207 \quad 0. = 1071 \text{ lead.}$$

and

$$0.2025 \div 0.1071 = 1.89.$$

The reaction between galena and silver-benzol sulphonate may therefore be written

$$PbS + 2(C_6H_5SO_3Ag) = Ag_2S + (C_6H_5SO_3)_2Pb.$$

The normal reaction occurring between galena and a soluble silver salt may therefore be regarded as strictly metathetical.

Mr. Palmer's experimental results are in agreement with the microscopic evidence in indicating that argentite is one of the stable products of the reaction of a soluble silver salt upon galena. The nature of the intermediate replacement product revealed by the microscope between galena and argentite still remains in doubt; it may be a solid solution of galena and argentite or a double sulphide of lead and silver. Nissen and Hoyt[1] from a study of a series of melts of galena and silver sulphide conclude that "the limit of solid solution at atmospheric temperatures is below 0.2 per cent Ag_2S." The apparent homogeneity of the transition mineral and the extreme sharpness of the boundaries between it and both galena and argentite seem to favor its interpretation as a double sulphide of lead and silver.

Although galena is the mineral most commonly replaced by argentite, a few instances of the replacement of polybasite by argentite or an association of argentite and chalcopyrite were noted, as shown in figure 11.

CHALCOPYRITE.

The copper-iron sulphide chalcopyrite has been formed rather commonly by replacement

[1] Nissen, A. E., and Hoyt, S. L., On the occurrence of silver in argentiferous galena ores: Econ. Geology, vol. 10, p. 179, 1915.

of galena, either alone or in association with argentite. Where galena has been replaced by a mixture of argentite and chalcopyrite, transition rims of the possible lead-silver sulphide are invariably present between argentite and galena but not between chalcopyrite and galena, a feature which is shown in figure 14. Replacement of polybasite by an association of chalcopyrite and argentite is shown in figure 11. In a very few places chalcopyrite was noted with polybasite in replacement vein-

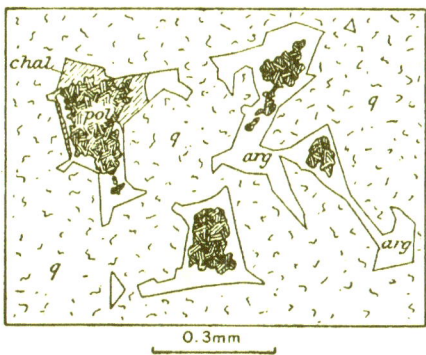

FIGURE 11.—Replacement of polybasite (pol), probably alpha hypogene, by argentite (arg) and some chalcopyrite (chal). q, Quartz. Camera lucida drawing from polished surface of ore from Favorite vein, 1,000-foot level of Tonopah Belmont mine.

lets traversing sphalerite, as is shown in Plate VIII, A, and figure 7.

CARBONATE-SULPHIDE AGGREGATES.

Fine aggregates of light-colored carbonate with argentite or polybasite and locally chalcopyrite, identical in character with those described above as probably representing alpha hypogene crystallization, replace galena in a number of the ores. Such replacement is clearly shown in Plate IX, A. Carbonate-sulphide aggregates replacing galena occur in the same specimen with aggregates of identical character that apparently have directly replaced the host rock.

Proofs that the aggregates replace galena are found (1) in the minutely irregular character of their contact with galena; (2) in the presence of small "outliers" of galena in the carbonate areas that have the same crystallographic orientation as the larger galena areas near by, of which they were evidently once a part; (3) in the complete transition of carbonate-polybasite areas into areas of polybasite that clearly replace galena, in the general fashion shown in figure 6; and (4) in the minor irregularity of the contacts between quartz and the carbonate-sulphide aggregates contrasted with the smoothness of those between quartz and galena (a contrast well shown at A and B in Pl. IX, A), indicating

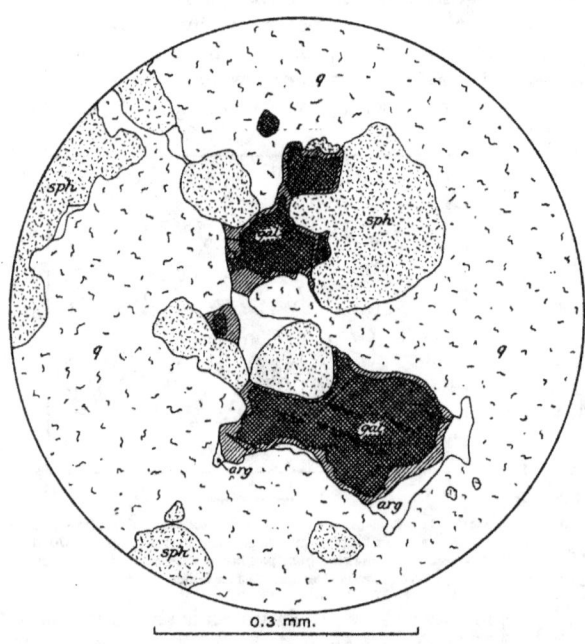

FIGURE 12.—Replacement borders of an undetermined mineral, probably a lead-bearing silver mineral, around galena (*gal*). The undetermined mineral is in turn replaced locally by argentite (*arg*). The triangular pits developed in the grinding of the ore surface are common to the galena and the bordering replacement rims. The surface was tarnished with hydrogen peroxide to differentiate galena from the undetermined mineral. *q*, Quartz; *sph*, sphalerite. Camera lucida drawing from polished surface of ore from Shaft.vein, 1,000-foot level of Tonopah Belmont mine.

that not galena alone but also small amounts of quartz have been replaced by the aggregate.

ELECTRUM.

Most of the electrum of the Tonopah ores appears to replace rock minerals (alpha hypogene) directly, as shown in Plate V, A and B; some of it is clearly formed by the replacement of galena. In Plate IX, B, is shown peripheral replacement of galena by an association of electrum and argentite. While argentite and electrum are sharply segregated in different parts of the replacement border, the galena boundary shows no jog in passing from the contact with argentite into the con-

tact with galena. Such replacement presents some analogies to the replacement of galena by the carbonate-sulphide aggregates shown in Plate IX, A, though in the electrum-bearing deposit only two replacing minerals are involved. Electrum is, indeed, a component of a few such carbonate-sulphide aggregates.

SUMMARY OF EVIDENCE OF HYPOGENE REPLACEMENT.

The conclusion that the replacement of one ore mineral by another in the Tonopah ores occurred mainly during the hypogene mineralization is perhaps the most important contribution of this paper. It is therefore desirable to summarize briefly the grounds upon which this conclusion rests.

1. The minerals that are described as formed by hypogene replacement of other ore minerals include no species that are not also found as undoubted alpha hypogene minerals—that is, directly replacing the host rock. The occurrence of electrum of identical appearance intergrown irregularly with sphalerite (alpha hypogene) and also peripherally replacing galena (beta hypogene) is particularly significant.

2. The fine carbonate-sulphide aggregates that replace galena (Pl. IX, A) are identical in mineral composition and in texture with others (Pls. VI, B, and VII, A) that directly replace the host rock and that are irregularly intergrown with coarser and undoubtedly hypogene sulphides.

3. Continuous areas of certain minerals and mineral aggregates were noted which appeared at one point to replace the host rock and at another point to replace earlier ore minerals. This was notably true of the carbonate-sulphide aggregates.

4. The presence of electrum peripherally replacing galena is highly suggestive of hypogene origin, for it is well known that gold and silver tend to part company in the processes involved in supergene enrichment because of their great difference in solubility under conditions of oxidation. If the electrum is hypogene the argentite which accom-

panies it in the peripheral replacement of galena (see Pl. IX, B) is also hypogene.

5. Replacement of the type believed to be hypogene is as conspicuous in unoxidized ore from some of the deepest workings (for example, in ore from the 1250-foot level of the Murray vein, Pl. IX, B; and from the 1200-foot level of the Belmont vein, figs. 7 and 9) as in ores near the surface. This argument is believed to have weight in spite of the great depth to which oxidation locally extends.

6. The prevalence of the supposed hypogene replacement minerals does not appear to be closely dependent upon the degree of fracturing, as it presumably would if the replacement had been accomplished by supergene solutions. The microscope shows that replacement has been very extensive in ores which exhibit little fracturing and no deposition of supergene minerals along the few fractures that are present.

SEQUENCE IN DEPOSITION AMONG HYPOGENE MINERALS.

The alpha and beta hypogene minerals, which, to simplify the discussion, have been separately described, are believed to have been

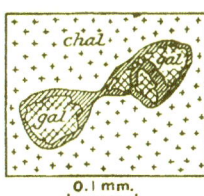

FIGURE 13.—Galena (gal) area in chalcopyrite (chal) partly replaced by an undetermined mineral, probably a lead-bearing silver mineral. Camera lucida drawing from polished surface of ore from West End vein, between 700 and 800 foot levels of Jim Butler mine.

deposited during a single period of mineralization, in the course of which the composition of the mineralizing solutions gradually changed. In the early stages of the process ore minerals were deposited mainly through replacement of the host rock; later, when less of the host rock remained near the vein fractures and when the character of the mineralizing solutions had changed, certain of the ore minerals, notably galena, were themselves replaced by other ore minerals.

The microscopic relations of the hypogene minerals show that in general quartz and sphalerite were deposited most abundantly early in the mineralization; galena, chalcopyrite, pyrargyrite, polybasite, argentite, electrum, and carbonates of calcium, magnesium, iron, and manganese formed most abundantly a little

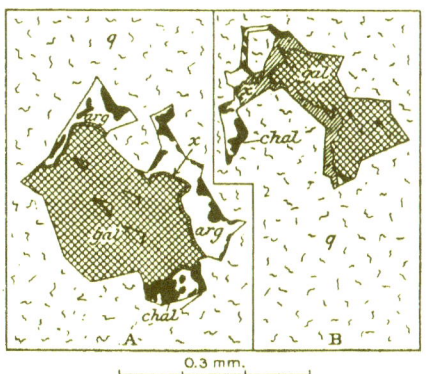

FIGURE 14.—Partial replacement of galena (gal) by an unidentified mineral, probably a lead-bearing silver sulphide (x); this in turn is replaced by an association of argentite (arg) and chalcopyrite (chal). In the left-hand figure the unidentified mineral forms only a narrow band between galena and argentite and is not present between galena and chalcopyrite. q, Quartz. Camera lucida drawings from polished surface of ore from West End vein, between 700 and 800 foot levels of Jim Butler mine.

later in the process. Still later galena and to some extent sphalerite, quartz, and polybasite became unstable and were replaced by one or more of the minerals argentite, chalcopyrite, carbonates, and electrum, whose deposition appears to have persisted nearly to the end of the hypogene mineralization.

Hypogene replacement of the type here described has not been widely recognized by students of ore deposits, because methods of microscopic study necessary to their perception have been developed only within recent years. The studies of Rogers[1] and Ray[2] on the copper ores of Butte, Mont., indicate that hypogene replacement has probably been a widespread process in that district, although the progress of the mineralization appears to have been punctuated by periods of fracturing to a greater extent than at Tonopah.

It appears certain that considerable changes in physico-chemical equilibrium, due to variations either in pressure or temperature or in the composition of the mineralizing solutions, have occurred during the formation of nearly

[1] Rogers, A. F., Upward secondary sulphide enrichment and chalcocite formation at Butte, Mont.: Econ. Geology, vol. 8, pp. 781-794, 1913.
[2] Ray, J. C., Paragenesis of the ore minerals in the Butte district, Mont.: Econ. Geology, vol. 9, pp. 463-481, 1914.

all ore deposits. In ores that are mainly fissure fillings these changes may be recorded by the presence of a suite of minerals in vugs and the medial portions of the veins different from those next the walls, or by crustification in which the layers formed last are mineralogically different from those formed first. In ores formed mainly by replacement such features as those observed at Tonopah may not be present if the earlier minerals are stable in the presence of the later solutions. The writers are inclined to believe, however, that careful study of other ores will show that replacement of earlier ore minerals by later ones during hypogene mineralization is not uncommon.

Although silver minerals and possibly gold were deposited in greatest abundance in the later stages of the hypogene mineralization at Tonopah, the writers prefer not to refer to the process as "upward enrichment," reserving that term for the rare occurrences in which the enriching metals can be shown to have been leached from deeper-lying ore by ascending solutions that were not at the start highly metalliferous. The deeper ores of Tonopah show no evidences of notable solution of any minerals except quartz. In the restricted sense indicated the term may find an application in discussions of certain western deposits in which thermal waters are now ascending along veins. The senior writer has now in hand the study of a number of such deposits.

NATURE OF DEEP MINE WATERS.

Warm waters have been encoutered in a number of the deeper mine workings at Tonopah, particularly in the western part of the camp. The questions arise: Are these waters now depositing ores? If not, what relation, if any, do they bear to the solutions that did deposit the ores?

Water from the Halifax mine, 1,400-foot level, about 380 feet east of the shaft, flowing as a small stream from a fracture in slightly mineralized Mizpah trachyte, had when collected a temperature of 106° F. (41° C.). This water was not acid toward methyl red solution and was not alkaline toward phenolphthalein. It was colorless, clear, and tasteless.

Water from the Tonopah Belmont mine, 1,500-foot level, about 100 feet north of the Belmont vein, had as it issued from the rock a temperature of 99° F. (37° C.). It flowed from

drill holes and small fractures in the face of a crosscut and was depositing a sludgy mass of hydrous oxides of iron (ocher-yellow) and oxides of manganese (black). The water as tested in the mine was not acid to methyl red solution and did not appear to be alkaline to phenolphthalein. Field tests with standard potassium permanganate solution showed that practically all the iron was in the ferrous condition. Two gallons of this water was collected and was analyzed by Chase Palmer, of the Geological Survey, with the results shown on page 29.

A sample of water from the Mizpah mine, collected by Spurr, was analyzed by R. C. Wells, of the Geological Survey, in 1910; the results are also recorded in the table. This water flowed from a drill hole which started from the 1,500-foot level of the Mizpah shaft and penetrated 816 feet below that level. Its temperature at the point of issuance was about 106° F.

The analyses are stated in the table in terms of radicles (or the roots of substances), such as Na, Ca, SO_4, and CO_3, which experience has shown to represent basal units of chemical interchange. This method of statement rather than a statement in terms of oxides is particularly appropriate to water analyses, for it indicates the points at which, according to the conceptions of the physical chemist the dissociation of the molecules of dissolved substances occurs. Most substances in water solution are not completely dissociated, the degree of dissociation differing for different substances and varying with temperature, pressure, the concentration of the solution, etc. In general the more dilute the solution the greater the proportion of ionization.

From the mere statement of the analysis in parts per million of the various radicles it is impossible to obtain a clear insight into the quality of the water, because each radicle has a different combining weight. It is convenient, therefore, to divide the total weight of each radicle present by the combining weight of that radicle, to obtain figures representing the reacting values of the radicles. The values so obtained are given in the columns headed "Reacting value by weight" and may be balanced directly one against another by simple addition or subtraction. The 2.205 reacting value of SO_4 in the Belmont water, for example, may be balanced against 3.472 reacting value of Na, leaving 1.267 (3.472 − 2.205) reacting value of Na to be balanced against other radicles.

Even this mode of statement is not fully satisfactory when it is desired to compare the qualities of several waters, for the reacting values obtained vary with the concentration of the solution. If the Belmont water, for example, were diluted with an equal volume of pure water the reacting value of each radicle would be halved. It will be readliy understood also that where dilute solutions act through considerable periods of time two waters carrying like constituents in like proportions but differing in concentration will in general produce similar chemical effects but at different rates. Although the concentration of the solution must therefore be duly considered, it is convenient in comparing waters to eliminate this factor. This result may be accomplished by rating the total value of all the radicles as 100 and proportioning the value for each radicle to this total; this has been done in the columns headed "Reacting value by per cent."

The method of statement and interpretation of water analyses here outlined is essentially that developed by Herman Stabler [1] and Chase Palmer,[2] of the United States Geological Survey.

The factors which in general exert the greatest influence on the potency of natural waters as geologic agents are the hydrogen ion and the hydroxyl ion. Hydrogen ions give acidity and hydroxyl ions alkalinity (or basicity); to the physical chemist, indeed, "hydrion" is the true acid and "hydroxion" the true base. Acid and alkaline waters are, at least so far as ore deposition is concerned, the most potent of natural solutions. Hydrogen ions may result from the ionization of free acid present in the water, as they possibly have in the acid water from the West End mine, Tonopah, described on page 43. (See analysis, p. 29.) Hydroxyl ions may result from the ionization of hydroxides present in the water, as in a deep water from the Comstock lode recently analyzed for the senior author by Mr. Palmer. Hydrogen ions may also be liberated less directly by the hydrolysis of various salts and the ionization of the resulting

acids; the acidity of copper sulphate in aqueous solution is attributable to this action. Similarly, hydroxyl ions may be liberated by the hydrolysis of various salts and the ionization of the resulting hydroxides; to this action is attributable the alkalinity of a solution of sodium carbonate.

The geologic importance of the presence of so-called free acid or free hydroxides in natural waters is self-evident. The geologic importance of the formation of hydrogen or hydroxyl by hydrolysis and ionization of salts might on first thought be questioned, for the proportion of a salt dissociated at one time is in most cases very small. As soon, however, as the small amounts of hydrogen or hydroxyl formed are neutralized, other hydrogen or hydroxyl ions are at once liberated, and by such progressive action a large part of a salt present in solution may eventually be involved in a reaction. A solution of copper sulphate carrying at any one time only a small proportion of hydrogen ions may, for example, react with calcium carbonate until a large part of the copper is converted into copper carbonate.

The amount of so-called free acid or free alkali present can be computed from the water analysis. At the present stage in the study of water analyses the extent to which hydrogen or hydroxyl ions will be formed through hydrolysis and ionization of salts must be inferred from a consideration of the relative strength of the acids and bases that compose the salts, on the principle that salts of strong acids and weak bases yield an excess of hydrogen ions giving acidity, salts of strong bases and weak acids yield an excess of hydroxyl ions giving alkalinity, and salts of strong bases and strong acids do not hydrolyze perceptibly, and give merely neutrality or salinity in the restricted sense of the term.

The considerations outlined above and the need of reducing so far as practicable the multiplicity of terms in most water analyses make it desirable to group the basic radicles in the general order of the relative strength of their hydroxides and the acid radicles in the general order of the relative strength of their acids or combinations with hydrogen. In the analyses in the accompanying table (p. 29) they are grouped as (1) strong base or alkali radicles, (2) weaker base or alkali-earth radicles, (3) weakest base or metal radicles, (4) strong acid

[1] Stabler, Herman, The mineral analysis of water for industrial purposes and its interpretation by the engineer: Eng. News, vol. 60, p. 356, 1908; The industrial application of water analyses: U. S. Geol. Survey Water-Supply Paper 274, pp. 165-181, 1911.

[2] Palmer, Chase, The geochemical interpretation of water analyses: U. S. Geol. Survey Bull. 479, 1911..

radicles, and (5) weak acid radicles. As the strength of an acid or base is, according to the ionic conception, a direct measure of the degree of ionization [1] such a grouping is a logical preliminary in any attempt to estimate the probable ionization in the solution. By balancing group totals of basic radicles against group totals of acid radicles in accordance with the well-established chemical principle of satisfying the strong acids and bases before the weak, an insight may be obtained into the quality of the water. In some waters individuals within the groups must also be considered. By applying this method of study it is generally possible to determine from the analysis with a fair degree of probability not only whether the water is acid, alkaline, or neutral, but also what constituents have conveyed these qualities to the water. Such computations should be supplemented in the field and in the laboratory by testing the water with indicators. In the present studies methyl red, methyl orange, and phenolphthalein were used in the field.

The details of the method of balancing the radicles are set forth below:

A. Balancing of strong acids.

1. Strong acid radicles should be balanced first against strong base or alkali radicles. Salts derived from strong acids and strong bases do not hydrolyze appreciably in water solution and so impart neither alkalinity nor acidity but merely salinity to the solution.

2. Remaining strong acid radicles should be balanced against alkali-earth radicles. Most salts of the alkali earths and strong acids do not hydrolyze appreciably in water solutions and so impart only salinity to the solution.

3. Remaining strong acid radicles should be balanced against the metallic radicles, the term metallic being here used in the restricted sense, exclusive of the alkali metals and metallic earths. Most of the salts of the metallic elements and strong acids hydrolyze to a considerable degree in water solution. The resulting strong acids ionize to a much greater extent than the weak hydroxides that are also formed, giving an excess of hydrogen ions over hydroxyl ions and rendering the solution acid. Ferric and copper sulphates in water solutions furnish excellent examples of acidity produced in this way. Waters rich in ferric sulphate will vigorously attack iron pipes in mines, even though sufficient iron is already present in solution to balance all the sulphate radicles.[2] The degree

of acidity imparted by such salts varies greatly, according to the relative strengths of the acids and bases concerned, so that each association must be individually considered.

4. Remaining strong acid radicles should be considered as in balance with hydrogen radicles—that is, as free strong acid.

B. Balancing of weak acids.

5. Weak acid radicles should be balanced first against alkali radicles if any remain after satisfying the strong acids. Most salts of the alkalies and weak acids hydrolyze to a notable degree in water solution. The resulting alkaline hydroxides ionize to a much greater degree than the weak acids also formed, giving an excess of hydroxyl over hydrogen ions and imparting alkalinity to the solution.

6. Remaining weak acid radicles should be balanced against remaining alkaline-earth radicles. Salts of the alkaline earths and weak acids hydrolyze to some extent in water solutions, yielding hydroxyl ions in excess of hydrogen ions and imparting alkalinity to the water.

7. Remaining weak acid radicles should be balanced against remaining metallic radicles (exclusive of the alkali or alkali-earth metals). The acids and hydroxides resulting from the hydrolysis of these salts are both weak; their ionization being slight, its effect is not likely to be pronounced in the direction of either acidity or alkalinity.

8. Remaining weak acid radicles should be considered as in balance with hydrogen radicles—that is, as free weak acid.

In practice there are always analytical errors that require adjustment in calculations of this sort. If the basic radicles are in excess the reacting values of all of them may be reduced proportionately until their total exactly equals that of the reacting values of the acid radicles. If the acid radicles are in excess and the water is not acid to indicators there has probably been an error in their determination, and the reacting values of all of them may be reduced proportionately until their total exactly equals that of the reacting values of the basic radicles. If the water is acid to indicators, a notable excess of acid over basic radicles is usually interpreted as indicating the presence of free acid; a slight excess is not significant, as the observed acidity may be due to hydrolysis and ionization of salts of strong acids and weak bases. If it is known that the analytical error probably lies in the determination of certain radicles it may be desirable to apply the correction to those radicles only.

[1] Walker, James, Introduction to physical chemistry, p. 313, 1913.
[2] Jones, L. J. W., Ferric sulphate in mine waters, and its action on metals: Colorado Sci. Soc. Proc., vol. 6, pp. 46–55, 1897.

Analyses of mine waters from Tonopah, Nev.

Radicles.	Belmont, 1,500-foot level, temperature 80° F.; Chase Palmer, analyst.			Mispah, 2,000-foot drill hole, temperature near 100° F.; R. C. Wells, analyst.			West End, 500-foot level; analyst Chase Palmer, analyst.		
	Parts per million.	Reacting value by weight.	Reacting value in per cent.	Parts per million.	Reacting value by weight.	Reacting value in per cent.	Parts per million.	Reacting value by weight.	Reacting value in per cent.
Na	80	3.422	31.4	148.8	6.473	30.2	135	5.859	12.7
K	5	.128	1.2	3.4	.087	.4	12	.307	.7
Total alkali radicles			32.6			30.6			13.4
Ca	20	1.000	9.1	68.8	3.433	16.0	249	12.450	27.1
Mg	4.4	.362	3.3	6.3	.518	2.4	10	.822	1.8
Total alkali earth radicles			12.4			18.6			28.9
Fe (ferrous)	3	.107	1.0	.7	.025	.1	7	.252	.5
Al				.7	.067	.4	3	.332	.7
Mn (bivalent)	4.12	.436	4.0				75	2.720	5.9
Zn				Trace					
(H)									(0)
Total metal radicles (Deficit of basic radicles)			5.0			.5			7.1
						1.5			
SO4	106	2.205	20.2	327.2	6.806	31.7	1,019	21.195	46.0
Cl	33	.932	8.0	169.1	1.001	4.7	65	1.833	4.0
Br and I				None			(e)		
NO3				Trace			None		
Total strong acid radicles			29.2			36.4			50.0
CO3	36	1.199	10.9	10.6	.353	1.6	None		
As3							Trace		
HCO3	51	.836	7.6	157.2	2.578	12.0	None		
Total weak acid radicles			18.5			13.6			
SiO2	18			64.8			15		
(Deficit of acid radicles)			12.3						
Total solids	367.4			821.1			1,560		
		10.732	97.7		21.354	99.7		45.750	99.4
		(f 11.010)	(100.0)		(g 21.432)	(100.0)		(g 46.056)	(100.0)

a The reacting values of the basic radicles in this analysis exceed those of the acid radicles but the discrepancy is within the permissible limit of experimental error.

b In this analysis the reacting values of the acid radicles exceed those of the basic radicles by a very slight margin, well within the permissible limits of experimental error.

c This water was found by testing in the field to be notably acid toward methyl red. The slight excess of acid radicles in the analysis is within the permissible limits of experimental error and is not therefore a certain indication of the presence of free acid. It is uncertain whether the acidity toward methyl red is due to free acid or to the hydrolysis and ionization of the small amounts of iron and aluminum salts present (ferrous and aluminum sulphates in aqueous solutions are both acid to methyl orange).

d At the time of analysis this water had deposited no sediment but on subsequent exposure to the air for several weeks all of the manganese deposited.

e Bromine and iodine if present are in minimal amounts.

f Twice the total reacting value of basic radicles.

g Twice the total reacting values of acid radicles.

Instead of distributing the error among all the acid or basic radicles it is in most analyses only necessary to distribute it among the group totals, as has been done for the first two analyses in the table. The results of the application of these methods to the two deep mine waters shown in the table are given in the following paragraphs:

Thermal water from Belmont mine, 1,500-foot level:

Alkali radicles balanced by strong acid radicles   58.4
Alkali radicles balanced by weak acid radicles..   6.8
Alkali-earth radicles balanced by weak acid
    radicles....................................   24.8
Metal radicles balanced by weak acid radicles..   10.0
                                                 ———
                                                 100.0

The above statement shows that this water should be alkaline because of the presence of alkali and alkali-earth carbonates which by hydrolysis and ionization give an excess of hydroxyl over hydrogen ions. Tests with phenolphthalein at the time of analysis gave an alkaline reaction. As noted on page 26, the water at the time of collecting did not appear to be alkaline toward phenolphthalein, but the test was made underground, and a faint pink coloration may have escaped detection. It is hardly probable that there was any appreciable absorption of alkali from the glass of the bottle in which the sample was shipped.

The water may be described as essentially a sodium and calcium sulphate and carbonate water which is somewhat alkaline and carries considerable manganese. In view of the absence of manganese in the Mizpah water it is not improbable that the Belmont water acquired its manganese content from manganiferous ores with which it came into contact.

Thermal water from Mizpah mine:

Alkali radicles balanced by strong acid radicles   61. 2
Alkali earth radicles balanced by strong acid radicles 11. 6
Alkali earth radicles balanced by weak acid radicles 25. 6
Metal radicles balanced by weak acid radicles..   1. 6
                                                     ------
                                                     100. 0

This is also essentially a sodium and calcium sulphate and carbonate water but has more than twice the concentration of the Belmont water and carries proportionately much more silica. No field tests with indicators are recorded by Spurr, but from the analysis this water would appear to possess only the very faint alkalinity attributable to alkali-earth bicarbonates. Unlike the Belmont water it carries no manganese. The presence of traces of nitrate in the Mizpah water is of interest in view of the known accumulation of nitrogen gas in the tops of certain stopes and raises at Tonopah.

Although both these waters are hot and the Mizpah water is clearly ascending from a considerable depth, there is nothing in their composition to indicate whether they come from an igneous source, or are surface waters that descended, became heated in some manner, and rose again toward the surface, or represent mixtures of waters from several sources.

No indications of the deposition of sulphides from these waters were found in the mines. As, according to the analyses, the waters contain no sulphur beyond that present in the sulphate radicle, there was probably no deposition of sulphides from them merely through a decrease in their pressure and temperature during their ascent. The deposition of sulphides and of certain native metals from sulphate solutions of this general character through reaction with earlier sulphides appears, however, to be possible. Such reactions may be metathetical, or they may involve reduction by the sulphides of certain metallic salts in the water. Iron and manganese, the only metallic components except aluminum sufficiently abundant to be determined in the water analyses would probably not be susceptible of such precipitation. The iron and manganese in these waters as they issued from the rock were already in a reduced condition incompatible with further reduction. As these metals stand at the end of the Schuermann series,[1] they are not likely to displace metathetically other metals from their combinations with sulphur. Certain other metals, however, notably silver, might if present be precipitated.

It may be concluded that these waters, if forming ore at all, are depositing sulphides or native metals only in small amounts by interaction with earlier sulphides.

It appears idle to speculate upon the possible relationship between these waters and the solutions that deposited the hypogene ores, either as regards origin or composition. They differ from those solutions at least in the absence of sulphides, for as the rocks that were replaced by the hypogene ores contained no minerals capable of reducing sulphates to sulphides, it is necessary to assume the presence of sulphides in the ore-depositing solutions.

SUPERGENE MINERALIZATION.

GENERAL CHARACTER.

Many of the hypogene ores have been extensively modified through the action of the air and of waters of surface origin descending through the lodes. The extent and degree of such action are dependent on the readiness with which the agents of alteration can gain access to the primary ores and are therefore intimately related to the nature and degree of frac-

[1] Schuermann, Ernst, Ueber die Verwandtschaft der Schwermetalle zum Schwefel: Liebig's Annalen, vol. 249, p. 326, 1888.

turing of the ores and wall rocks and the nature of the ground-water circulation. The complex history of the superficial alterations at Tonopah, involving two or more periods of oxidation subsequent to the primary mineralization, makes it unusually difficult to classify the alteration products in close accordance with the physical and chemical conditions under which they were formed. In the descriptions that

west of the fault plane proper the ore is much shattered and is considerably more oxidized than the less shattered ore farther west. Slight oxidation near fractures was noted in ore of the Belmont vein on the 1,500-foot level of the Tonopah Belmont mine. In the workings on the Murray vein slight oxidation was noted on the 1,170-foot level but not on the 1,260-foot or deepest level.

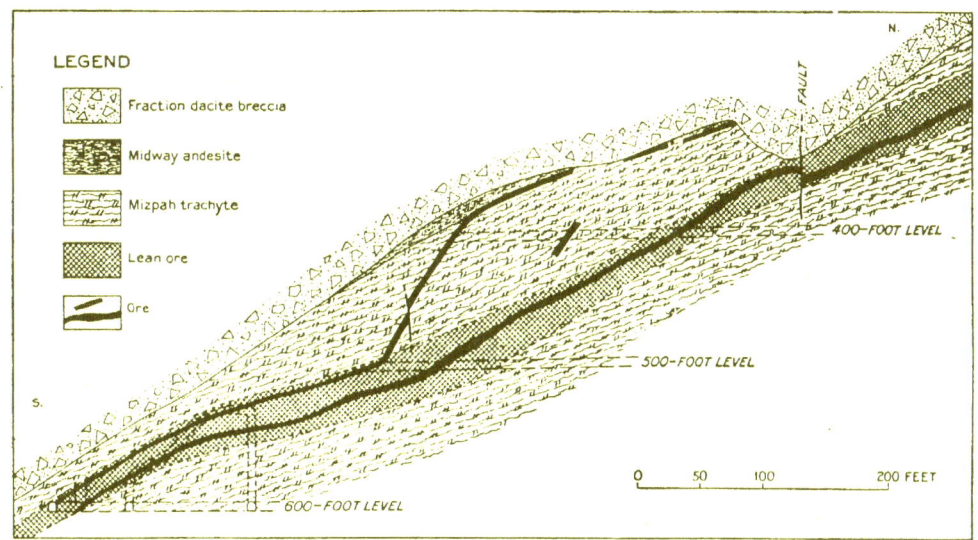

FIGURE 15.—Section through portion of West End mine showing structural relations of the Fraction dacite breccia to the West End vein and to the Mizpah trachyte.

follow they have been grouped according to their mode of occurrence as (1) residual products of oxidation and deposits in open spaces and (2) supergene replacement deposits.

DEPTH OF OXIDATION.

General oxidation is characteristic only of the upper portions of the few veins that crop out. In the upper portions of the veins that do not crop out and in the deeper portions of the outcropping veins oxidation is local only. Such local oxidation was noted in some of the deepest workings near cross fractures that intersect the veins. It is well shown, for example, on the 770-foot level of the Tonopah Extension mine, the deepest level reached by the workings connected with the main shaft, where fractured ore and wall rock just above the Rainbow fault are much more oxidized than those at a distance from the fault. On the 600-foot level of this mine the Extension north vein is cut and offset by the Rainbow fault; for over 100 feet

OXIDATION AT SEVERAL PERIODS.

At a number of localities in Tonopah there is evidence of oxidation connected not with the present surface but with ancient land surfaces now deeply buried. The most striking evidence of this sort is found at several points near the 300 and 400 foot levels of the West End mine, where the ore of the West End vein is capped for short spaces by the Fraction dacite breccia. In the vicinity of the West End mine the Fraction dacite breccia is the surface rock; its base appears to have a general southwesterly dip of about 30° and truncates the Midway andesite, the Mizpah trachyte, and in a few places the West End vein. The general structural relations are shown in figure 15.

The dacite breccia is, at least in large part, a volcanic tuff, which, like the other volcanic rocks of the region, is believed by Spurr to be of Tertiary age. Its tuffaceous nature is suggested by its appearance to the unaided eye but is especially evident under the microscope, as shown in Plate X, A (p. 25). The specimen studied microscopically and figured in this illustration was obtained near station 1447 on the 300-foot level of the West End mine. It contains rock fragments irregularly angular in form and of various sizes and kinds; most of them are so much altered that their original character can not be determined; some show plagioclase phenocrysts in a finely granular ground mass. In addition, there are numerous fragments of quartz and feldspar phenocrysts, as shown in the illustration. The feldspar crystals are mostly andesine. A few of the feldspar crystals are perfect, but most of them are mere fragments, and many are very angular and irregular in form. Some are much fractured, not along cleavage

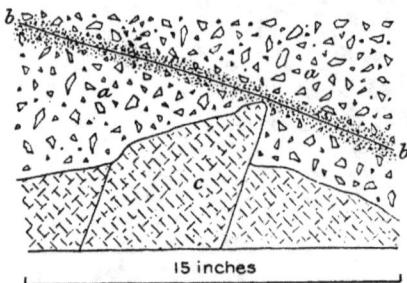

15 inches

FIGURE 16.—Sketch of a portion of the contact between the Fraction dacite breccia and the West End vein in stope 309, West End mine. *a*, Unoxidized breccia; *b*, breccia oxidized near fracture; *c*, highly oxidized ore.

planes but very irregularly; the fractures do not extend into the inclosing matrix. One feldspar crystal may be much fractured and another very close to it may be unfractured, indicating that the fracturing took place before the crystals became embedded in their present matrix. Under high magnification (240 diameters) the matrix still appears fragmental and the fragments show diversity of size and shape. The smallest fragments are isotropic and are possibly volcanic glass. The alterations include partial calcitization of the plagioclase, some secondary enlargement of quartz fragments, and the development of chlorite in certain rock fragments.

That the breccia is younger than the ore of the West End vein is shown by the presence of a number of fragments of ore and pyritized wall rock in the breccia near its contacts with the vein.

Where the relations are not complicated by faulting the breccia appears to have been deposited directly upon the ore; in places it shows an obscure stratification parallel to the contact. Oxidation of the breccia near its contact with the vein is not general, but is usually confined to the vicinity of fractures, as shown in figure 16. Breccia in immediate contact with the vein may be wholly free from oxides of iron. The ore underlying the breccia, on the other hand, is as highly oxidized as much of the ore found at the present surface and carries iron oxides in mammillary and botryoidal forms. Some of the quartzose ore next to the contact is porous from partial solution of its constituents.

The high degree of oxidation in the ore, the unoxidized condition of much of the breccia in contact with the ore, and the local inclusion of oxidized fragments of ore in unoxidized breccia indicate that much of the oxidation of this ore took place before the deposition of the breccia.

Other evidence leading to the same conclusion consists in the presence of coatings of calcite, barite, crystalline quartz, and, rarely, pyrite on the hydrous iron oxides of the ore just below the breccia contact. (See Pl. X, *B*.) In ore from stope 484, not far above the 400-foot level of the West End mine, a coating of crystals of a carbonate of iron and calcium is deposited on hydrous iron oxides. Some barite is intergrown with the carbonate. On top of the carbonate occur well-formed crystals of clear quartz. Locally, quartz has replaced the exterior of scalenohedral crystals of calcite which rest on iron oxide. Ore from stope 400A shows thin coatings of pyrite on hydrous oxides of iron. These associations, except that of quartz, constitute mineral misalliances, for the minerals of the coatings are not normally formed during the oxidation of a sulphide ore. Their presence is, therefore, indicative of a change in the physical and chemical conditions surrounding this ore. The structural relations described on page 31 indicate that this change was brought about by the burial of an ancient, probably Tertiary gossan under a great thickness of volcanic tuffs, inducing a return from surficial to deeper-seated conditions.

Spurr [1] calls attention to the probability of a still earlier period of denudation and oxidation between the deposition of the Mizpah trachyte and that of the Midway andesite. He says: "Most likely the earlier andesite [Mizpah trachyte] was deeply eroded and the veins were exposed before the later andesite [Midway andesite] was poured out." The contact between these two formations is in most exposures a fault contact, so that conclusive evidence of oxidation during this period could not be obtained.

From the evidence above set forth it may be concluded that although a part of the oxidation, even in those veins which do not crop out, is certainly recent, some is to be correlated with ancient, probably Tertiary land surfaces. As downward enrichment has been an accompaniment of the recent oxidation, it is logical to suppose that it was also an accompaniment of the ancient oxidation, and that there have, therefore, been several periods of downward enrichment.

GROUND WATER.

The common text-book diagram in which the water table is shown as a subdued counterpart of the land surface is applicable in a general way to regions, such as the Mississippi Valley, characterized by moderate humidity and flat-lying strata through which lateral circulation of ground water is easy, but it does not at all express the relations at Tonopah. In the Tonopah mines the active ground-water circulation is practically confined to zones of fracturing; copious flows are limited to a few such zones and usually to their more open parts. Other fractures have, for considerable periods at least, and to great depths, been sufficiently free from water to permit extensive oxidation through the agency of the atmosphere. These relations were emphasized by Spurr,[2] who says:

No general body of ground water has been encountered, although the rocks are extremely fractured; yet along certain steeply inclined fracture zones water is found sometimes quite near the surface and occasionally in considerable quantity. This water is cool, is sufficiently nonmineral to be fair drinking water, and is undoubtedly the storage of precipitation.

[1] Spurr, J. E., Geology of the Tonopah mining district, Nev.: U. S. Geol. Survey Prof. Paper 42, p. 35, 1905.
[2] Idem, p. 107.

Spurr cites much evidence found in the sinking of shafts and driving of drifts illustrating the localization of the ground water in fractures. Observations from later mining experience, reported to the writers are wholly similar in their import and need not be enumerated here.

RESIDUAL PRODUCTS OF OXIDATION AND DEPOSITS IN OPEN SPACES BY SUPERGENE SOLUTIONS.

MINERALS NOTED.

The minerals listed below were noted either as residual untransported products of oxidation or as deposits in open spaces at a very late stage in the mineralization. As pointed out on page 42, certain of them, notably the sulphides, were formed as a rule at greater depths than others, but the fact that there have been several periods of oxidation makes it impracticable to define closely the conditions of formation of each mineral.

Native elements:	Oxides—Continued.
Gold.	Hydrous oxides of iron.
Silver.	Oxides of manganese.
Sulphides:	Carbonates:
Pyrite.	Calcite.
Chalcopyrite.	Mixed carbonates.
Argentite.	Malachite.
Sulpho-salts:	Silicates:
Polybasite.	Calamine.
Pyrargyrite.	Kaolin.
Haloids:	Phosphate:
Cerargyrite.	Dahllite.
Embolite.	Sulphates:
Iodobromite(?).	Barite.
Iodyrite.	Gypsum.
Oxides:	
Quartz.	
Hyaline silica.	

GOLD.

Grains of bright-yellow gold large enough to be seen with the unaided eye are commonly confined to the much oxidized ores of the few veins that crop out. Such gold was abundant at a depth of about 100 feet in the Valley View vein, above and below a fault that displaces the vein. In specimens examined by the writers from the 500-foot level on this vein it forms minute flakes on oxidized surfaces, on which iodyrite and iodobromite were also deposited. Gold in association with iodobromite was also noted in abundance as golden-yellow grains 0.1 to 0.2 millimeter in diameter in the pannings from soft oxidized ore from the South vein of the Desert Queen mine about 112 feet above the 600-foot level. Fractures in this ore are heavily coated with oxides of manganese.

The gold of the oxidized ores is commonly in larger grains than the hypogene electrum and, to judge from its yellow color, is of much greater purity. It is probably of supergene origin.

SILVER.

Silver occurs here and there in certain highly or moderately oxidized ores and in a very few places is fairly abundant. Argentite is in many places closely associated with it.

In the Tonopah Belmont mine native silver is present locally in the Belmont and Shaft veins. In the Belmont it was found from the 900-foot level, shortly below the "cap rock," down to the 1,400-foot level, though most abundant in the richest parts of the vein near the 1,100 and 1,166 foot levels; in the Shaft vein it was noted in ore from depths as far down as the 1,400-foot level. Silver was observed by the writers in ores from the MacDonald vein on the 330-foot level of the North Star mine and from vertical depths of 500 to 600 feet in the West End vein.

The silver may occur in wires, flakes, or tapering, fanglike forms commonly curling at the tip. The silver wires are found as a rule in matted aggregates in small vugs or open fractures, where they may be attached to quartz or barite crystals or to argentite. Ragged incrustations of silver on fractures and in small solution cavities were noted in ore from the 1,400-foot level on the Belmont vein, where they coat quartz or hydrous oxides of iron.

Small, bright flakes of silver of irregular outline on fractures in hypogene ore were obtained between the 1,300 and 1,400 foot levels of the Belmont vein were not evenly distributed over the walls of the fractures but were deposited only where hypogene sulphides were abundant, suggesting that the sulphides exerted an influence in the silver precipitation. In some slightly oxidized ores of the Belmont vein the silver flakes are associated with a soft claylike material of blue-green color. Close study of this material shows that it is probably kaolin colored by an adsorbed soluble salt of copper. When pulverized and treated with cold aqua ammonia the blue material imparts the charac-

teristic copper-blue to the solution. Acidification with hydrochloric acid yields no precipitate, showing that, contrary to the prevailing opinion among the miners, the blue material does not contain silver haloids. In this material are embedded flakes of silver and some grains of argentite, but crushing and panning may be necessary before they become visible. Local shearing since the silver was deposited has compacted the blue-green kaolin-like material into a talclike mass and drawn out the silver flakes into a very thin natural foil. Such shearing must be very recent.

Tapering, fanglike forms of silver were noted in ores from the Belmont, Shaft, and West End veins and commonly rise abruptly from a base of argentite or more rarely of polybasite; in some ores minute masses of argentite are attached to their sides or tips. Rarely the "teeth" are attached to a base of quartz or of barite. It is believed that the silver "teeth" have resulted from the reduction of silver compounds in place.

PYRITE.

Deposits of pyrite were noted in a few places coating fractures in hypogene ores to depths of about 600 feet. At several localities it is intergrown with barite, clear crystals of which in places completely inclose small pyrite pyritohedrons. Pyrite closely associated with barite was noted in fractures in the McNamara vein not far above the 300-foot level (344 raise) of the Tonopah Extension mine; in the Murray vein on the 600-foot level (end of 603 north crosscut) of the same mine; and in the West End vein on the 500-foot level (502 drift) of the West End mine. On the pyrite and barite locally ferruginous and manganiferous calcite, argentite, or both, have been deposited. In one place intergrown pyrite and quartz are deposited on earlier, more coarsely crystallized quartz coating a fracture in hypogene ore.

In all the late pyrite deposited in fractures in the Tonopah ores the predominant crystal faces are those of the pyritohedron, whereas in the pyrite of the undoubted hypogene ores and in the pyrite developed hydrothermally in the wall rocks cube faces predominate. That this contrast in crystal form signifies different conditions of deposition can hardly be questioned, but sufficient data are not at hand to warrant conclusions as to its precise significance.

CHALCOPYRITE.

Small amounts of very finely crystalline chalcopyrite occur as late deposits along fractures in hypogene ore. In the Murray vein not far below the 1,000-foot level (203 stope) of the Tonopah Extension mine, chalcopyrite, polybasite, barite, and argentite have been deposited on crystalline quartz that coats fractures in hypogene ore. The sulphides are locally on or in the barite; elsewhere barite crystals lie on the sulphides. A part of the argentite in these fractures is an alteration from polybasite, for the exterior portions of flat hexagonal crystals of the typical polybasite form are now sectile argentite.

In the specimen from the MacDonald vein illustrated in Plate XII, A, finely crystalline chalcopyrite locally coats the polybasite in fractures only partly filled by polybasite. In the Fraction vein between the 600 and 700 foot levels of the Jim Butler mine argentite, locally intergrown with barite, coats fractures in hypogene ore; and on the argentite or intercrystallized with it are small crystals of chalcopyrite. Ore from the West End vein between the 700 and 800 foot levels shows intimate mixtures of argentite and chalcopyrite coating fractures in hypogene ore, also chalcopyrite, polybasite, and pyrargyrite in quartz-lined vugs.

ARGENTITE.

Argentite is particularly common as a late deposit on the walls of fractures cutting the hypogene ores. In one specimen from the Jim Butler mine it partly fills fractures in clean white quartz in much the fashion that polybasite does in the specimen shown in Plate XII, A. A branching deposit of argentite on a fracture in quartzose ore of the MacDonald vein is shown in Plate XI, A, and lichen-like deposits from the Murray vein are shown in Plate XI, B. These deposits lie directly on the hypogene ore; in other places a thin layer of crystalline quartz intervenes.

In many fractures argentite is accompanied by light-colored carbonates. In some places the argentite coats the carbonate, as in the Fraction vein between the 600 and 700 foot levels of the Jim Butler mine (stope 4), where siderite on a fracture is coated with a thin deposit of argentite about the area of a silver dollar. In other places argentite and carbonate are intercrystallized in a fashion suggesting contemporaneity. In fractures in the West End vein—for example, in a specimen taken between the 400 and 500 foot levels of the West End mine—argentite has been deposited over some crystals of ferruginous calcite; elsewhere on the same specimen the calcite coats the argentite, or the two are irregularly intercrystallized. In the Fraction vein between the 600 and 700 foot levels of the Jim Butler mine the the wall of a small fracture cutting hypogene ore is coated with slightly ferruginous calcite; over this in places is an intimate mixture of calcite, argentite, and some barite, the argentite in part inclosed by the calcite or barite; and on this mixture pure argentite is locally deposited. In the West End vein, between the 400 and 500 foot levels of the West End mine; the Fraction vein, not far above the 700-foot level of the Jim Butler mine; and the Murray vein in the 1,000-foot level of the Tonopah Extension mine argentite was noted entirely inclosed by clear barite crystals.

POLYBASITE.

Polybasite is common as a late deposit in fractures traversing hypogene ores; in some places it is intimately intergrown with pyrargyrite, and in many others it is closely associated with argentite. The first coating on such fractures is commonly crystalline quartz, the polybasite being deposited on this quartz.

In the Belmont vein on the 1,000-foot level of the Tonopah Belmont mine, near south crosscut 1044, fractures cutting hypogene ore are lined with finely crystalline quartz, on which thin patches of polybasite and argentite are deposited. A particularly fine specimen showing polybasite deposited along fractures in white quartz from the MacDonald vein is shown in Plate XII, A. In one part of this specimen a little native silver and in another part finely crystalline chalcopyrite were deposited on the polybasite.

Polybasite in fractures is locally associated with light-colored carbonates. In the Murray vein not far above the 1,100-foot level polybasite in some places lies on calcite and in others is coated by calcite, the two minerals being essentially contemporaneous.

PYRARGYRITE.

Late pyrargyrite that can be cleanly stripped from fractures and vugs in hypogene ores was noted in several places but was much less abundant than the pyrargyrite described on pages 14–15, which is believed to be hypogene. Such late pyrargyrite is also less abundant in general than late argentite and polybasite.

Pyrargyrite with intergrown polybasite and chalcopyrite was noted coating quartz crystals lining vugs in the West End vein between the 700 and 800 foot levels of the Jim Butler mine. The sulphides could be cleanly stripped from the quartz and were plainly deposited at a later period. Branching forms of late pyrargyrite intergrown with some polybasite on the wall of a fracture in white quartz in the Montana Tonopah mine are shown in Plate XII, B.

In ore rich in rhodochrosite from the West End vein between the 400 and 500 foot levels of the West End mine pyrargyrite and polybasite form thin coatings on the walls of fractures, from which they can be cleanly stripped. Specimens showing identical relations were obtained from the Tonopah Extension office, but their original location in the Extension workings is unknown.

SILVER HALOIDS.

Silver haloids were noted only in the veins that crop out. They were reported from several parts of the Tonopah Belmont mine. Specimens said to contain them were found to carry native silver and a soft blue-green material, probably kaolin, with adsorbed soluble copper salts, but no silver haloids were found. According to Burgess,[1] who had unusual opportunities for studying the occurrence of the silver haloids at Tonopah, the most abundant of them is cerargyrite, although most of the samples tested by him showed traces of bromine. All the gray silver haloids tested by the writers contained considerable bromine, but comparatively few specimens were available. There are evidently all gradations from cerargyrite into the chlorobromide embolite. Silver haloids, carrying chlorine, bromine, and iodine and others carrying bromine and iodine, were also noted. The pure bromide, bromyrite, was not noted,

but nearly pure iodyrite is locally abundant. The colors of the silver haloids are in places deceptive, for much translucent pale-gray haloid appears to carry as much bromine as chlorine, and some pale-gray iodyrite was noted.

The mode of occurrence of these minerals presents a number of interesting features. In one specimen from the Tonopah Mining Co.'s upper workings (exact locality unknown) fractures in sulphide ore were coated with hydrous oxides of iron; on these was deposited a thin coating of malachite, which, in turn, was coated with calcite. Cerargyrite of dirty-gray color, carrying some bromine, is locally intercrystallized with the malachite and calcite. In one place the cerargyrite is wholly inclosed in calcite, but a short distance away it lies on unetched crystal faces of calcite.

A "sand" from porous oxidized ore from the Mizpah vein between the 400 and 500 foot levels carries yellowish iodyrite and a yellowish-green haloid without cleavage, that appears, upon testing,[2] to contain much bromine and less iodine and chlorine. The associated iodyrite is pale yellow, reacts only for iodine, and shows the hexagonal form and basal cleavage characteristic of that mineral.

In other specimens from the Tonopah Mining Co.'s upper workings (exact locality unknown) iodyrite that was deposited on the surface of iron oxide coating quartz is in part pale yellow and in part dirty gray in color. Both varieties are hexagonal, have perfect basal cleavage and yellow streak, and react only for iodine when tested in the closed tube. Some crystals are gray iodyrite at the base and yellow above. In parts of one specimen a first generation of iodyrite crystals has been coated with transparent colorless hyaline silica, and on top of this a later generation of smaller iodyrite crystals has been deposited. A centimeter or so away on the same surface, where hyaline silica is not present, there is only one set of iodyrite crystals, the later iodyrite appearing to have enlarged the earlier crystals without forming new ones. Locally iodyrite crystals are entirely inclosed by the hyaline silica, show-

[1] Burgess, J. A., The halogen salts of silver and associated minerals at Tonopah, Nev.: Econ. Geology, vol. 6, p. 17, 1911.

[2] The silver haloids were mixed with a little pulverized galena and heated before the blowpipe in the closed tube in accordance with the method recommended by Brush and Penfield (Manual of determinative mineralogy, p. 68, 1911). The lead chloride, bromide, and iodide formed are deposited at different heights in the tube and are easily distinguished from one another. Their relative amounts can be roughly estimated.

ing that the two minerals formed contemporaneously. Another specimen from the same mine shows hyaline silica deposited on iodyrite and iodobromite deposited on top of the silica; still another shows both iodyrite and iodobromite on silica that coats earlier iodyrite. The largest crystals of iodyrite seen were in a vug in oxidized ore, in a specimen from the Tonopah Mining Co.'s property, in the collection of J. A. Uhland. They measured three-sixteenths of an inch across the hexagon.

The relation of the silver haloids to the phosphate dahllite is considered on page 38.

QUARTZ.

Quartz is abundant as a coating on the walls of fractures traversing hypogene ores. It commonly forms continuous coatings from 1 to 3 millimeters thick composed of minute colorless, transparent hexagonal crystals. It is usually the first coating on such fractures, and on it barite, carbonates, kaolin, polybasite, argentite, pyrite, chalcopyrite, and a few other minerals are locally deposited. These quartz coatings are possibly late hypogene, as noted on page 18.

HYALINE SILICA.

A number of characteristic occurrences of hyaline silica are described in the sections on silver haloids and dahllite. The mineral was also noted coating psilomelane in the Valley View vein on the 400-foot level. It is not confined to the veins but occurs abundantly at the surface as thin coatings along joints in the Midway andesite.

OXIDES OF IRON AND MANGANESE.

The presence of rhodochrosite in the hypogene ores has led to an abundance of hydrous oxides of manganese in the products of their oxidation. These are rarely pure, being mixed usually with hydrous oxides of iron. In the Belmont vein between the 1,300 and 1,400 foot levels there are masses a foot or so across composed almost wholly of these oxides. Thin exterior coatings of dark reddish-brown pulverulent oxides coat a porous framework of hard gray metallic oxides that give an ocher-yellow streak. On the 1,300-foot level on the same vein a lens of pulverulent manganese oxide 10 inches wide and 2 feet long was noted

at one place. The material is dark dull gray and finely crystalline, gives a dark-brown streak, and is probably manganite.

CALCITE.

Calcium carbonate nearly free from iron, manganese, and magnesium is much less common than the mixed carbonates described below and was noted in only a few places. Colorless calcite carrying a trace of manganese was noted in the Belmont vein on the 800-foot level of the Tonopah Belmont mine as a crystalline coating on fractures traversing coarse hypogene sulphide ore. In places this calcite is intergrown with crystals of polybasite, the two being apparently contemporaneous. Calcite was also noted coating malachite in ore from the Tonopah Mining Co.'s property.

MIXED CARBONATES.

Qualitative tests on many of the carbonate coatings on fractures in hypogene ore show the presence of two or more of the elements iron, manganese, calcium, and magnesium. Many carbonates that are pure white in color are notably ferruginous and turn brown on heating. Manganese in many places accompanies the iron, but its proportion appears less than in most of the hypogene carbonates, for a pink color was nowhere noted.

MALACHITE.

Carbonates of copper are not at all common in the ores now visible at Tonopah. In a specimen from the property of the Tonopah Mining Co. (exact location unknown) a fracture in the hypogene ore is coated with hydrous iron oxides; on these locally malachite and gray cerargyrite carrying some bromine have been deposited. Both these minerals are in places coated with transparent colorless calcite.

CALAMINE.

Characteristic "sheafs" of the hydrous zinc silicate calamine were noted in the West End vein between the 500 and 600 foot levels (stope 600 C) of the West End mine. A fracture in the sulphide ore was first coated with a thin layer of crystalline quartz; on this argentite and siderite were deposited; oxidation followed, converting part of the siderite to hydrous iron oxides and rendering the argentite dull and locally pulverulent; and finally calamine was deposited on the iron oxide.

KAOLINITE.

A white pulverulent infusible material that reacts strongly for alumina but not for potassium [1] is common in the vugs and pores of many of the ores. Microscopic examination shows that this material is an aggregate of minute, roughly spindle-shaped crystals of quartz and another mineral of nearly the same index of refraction but of very low double refraction. This second mineral, which is probably kaolinite, is in too minute grains to permit the determination of its optical constants.

Kaolinite has been one of the latest minerals to form in the Tonopah ores. Locally it occupies spaces formed by the solution of hypogene quartz or carbonates; elsewhere it has been deposited on argentite, polybasite, and carbonates that appear to be of supergene origin. In a specimen from a vertical depth of 1,395 feet on the Shaft vein (east stope 3) in the Tonopah Belmont mine kaolinite has been deposited subsequent to the formation of wires of native silver, which it locally coats. Its presence was noted in ores from the deepest workings; it is common in unoxidized ores; and where it occurs in oxidized ores its development has preceded oxidation except where, as shown by other evidences, there has been a return from oxidizing to deeper-seated conditions.

Kaolinite is not confined to the ores but is common in the neighboring wall rocks. In the north crosscut on the 850-foot level of the North Star mine, for example, it forms the matrix of rock fragments in a brecciated zone in the Mizpah trachyte.

Much of the kaolinite occurring in the ores is not a purely residual product but is either a precipitate from solution or has been introduced in suspension. This is shown by its mode of occurrence in certain quartz-lined vugs, where it is heaped up like snow on a fence top on the upper sides of the quartz crystals and is relatively scarce on their under sides.

DAHLLITE.

The presence of dahllite, the calcium carbono-phosphate ($3 Ca_3(PO_4)_2.CaCO_3$), at Tonopah was first recognized by Rogers [2] in ore from the Mizpah mine. This ore also contained iodyrite, hyaline silica, quartz, and manganese dioxide. The dahllite formed small white tabular crystals of hexagonal outline resembling apatite in form but effervescing with warm nitric acid.

The same mineral was observed by the writers in ore from the Valley View vein just below the 400-foot level. It forms short hexagonal prisms 0.5 millimeter or less in diameter that effervesce slightly with cold nitric acid and react for phosphorus with ammonium molybdate. The dahllite crystals lie upon quartz crystals lining vugs and also upon hydrous oxides of iron, and on the dahllite in turn silver haloids are deposited. All these minerals are covered locally by clear, colorless hyaline silica. Of the silver haloids olive-green iodobromite was first deposited, and on this and also on the dahllite is iodyrite.

In another ore specimen from the Tonopah Mining Co.'s workings (exact locality unknown) dahllite occurs both in and on hyaline silica coating oxidized surfaces.

It is apparent that the dahllite was deposited under the same conditions as the silver haloids and hyaline silica.

BARITE.

Barium sulphate is fairly abundant in well-formed crystals on fractures traversing hypogene ore; commonly the first coating is quartz and the barite crystals lie on that. In many places the barite was clearly deposited contemporaneously with one or more of the minerals pyrite, argentite, and polybasite, grains of these sulphides being wholly inclosed by the barite. In other places barite is the only coating on the fractures or on the quartz lining the fractures.

GYPSUM.

Selenite crystals are not uncommon in the oxidized ores, especially in the veins that crop out. Abundant selenite was observed in the

---

[1] The material was tested for alumina by moistening with cobalt nitrate and heating before blowpipe; for potassium by the flame coloration test.

[2] Rogers, A. F., Dahllite (podolite) from Tonopah, Nev.: Am. Jour. Sci., 4th ser., vol. 33, pp. 475–482, 1912.

South vein of the Desert Queen mine on the 700-foot level, where it occupies fractures in oxidized ore. Pyrolusite is the most abundant oxidation product there, and the selenite lies on the pyrolusite, being clearly of later formation.

EPSOMITE.

Crystals of hydrous magnesium sulphate were noted in the West End mine on the 400-foot level (near raise 415 B), where they had been deposited recently near cracks on the walls of the drift by the evaporation of water used in wetting down the stopes. The epsomite may have been an original constituent in the water supply or may have been acquired by the water through reaction with the waste in the stopes.

SUPERGENE REPLACEMENT DEPOSITS.

REPLACEMENT OF CHALCOPYRITE BY BORNITE (?), COVELLITE, AND ARGENTITE.

In the processes of mineralization thus far described chalcopyrite has replaced other minerals but has not itself been replaced. In other words, chalcopyrite has behaved as a stable mineral under the conditions of supposed hypogene mineralization.

In an ore specimen from the Shaft vein on the 1,000-foot level of the Tonopah Belmont

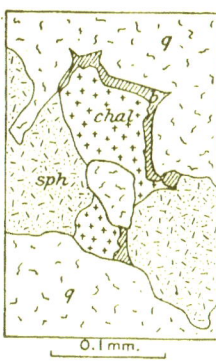

FIGURE 17.—Replacement rim of probable bornite (b) bordering chalcopyrite (chal). q, Quartz; sph, sphalerite. Camera lucida drawing from polished surface of ore from Shaft vein, 1,000-foot level of Tonopah Belmont mine.

mine, examined under the microscope, chalcopyrite has been replaced by what is probably bornite and this in turn by covellite, argentite, or an intergrowth of these two minerals, as shown in figures 17 and 18.

The supposed bornite rims around chalcopyrite showed the purplish color usual in freshly polished bornite. The quantities being microscopic, it was impossible to isolate any of the minerals for analysis. It is known that chalcopyrite immersed in silver sulphate solution soon becomes coated with a tarnish resembling bornite, probably a sulphide of silver and iron, as Palmer[1] has shown. This mineral is of course

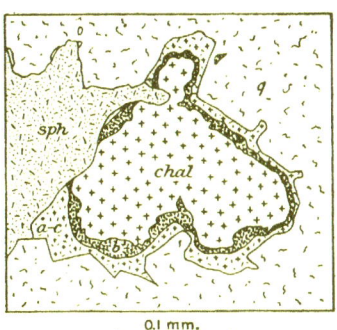

FIGURE 18.—Peripheral replacement of chalcopyrite (chal) by probable bornite (b) which is in turn partly replaced by a fine intergrowth of argentite and covellite (a-c). q, Quartz; sph, sphalerite. Camera lucida drawing from polished surface of ore from Shaft vein, 1,000-foot level of Tonopah Belmont mine.

stable in contact with silver sulphate solution. To determine whether the bornite-like rims were this mineral, they were treated with silver sulphate solution. They tarnished in the same manner as known bornite from Virgilina, Va., and are believed to be bornite. The covellite shows the characteristic blue color. There are gradations from apparently pure covellite through fine intergrowths of covellite and supposed argentite into areas that are unquestionably pure sectile argentite. When the fine intergrowths were treated with silver nitrate solution the covellite became gray and indistinguishable in tint from the supposed argentite with which it was intergrown. As it is well known that covellite changes to argentite on treatment with silver nitrate, the identification of both minerals appears reasonably certain.

The close association of covellite and argentite is of interest in view of the known capacity of covellite to form silver sulphide (Ag_2S) when treated with silver nitrate or silver sul-

[1] Palmer, Chase. Bornite as silver precipitant: Washington Acad. Sci. Jour., vol. 5, pp. 351–354, 1915.

phate solution.[1] In this specimen the argentite appears not to be a replacement of the covellite but to be contemporaneous with it. Covellite alone peripherally replacing chalcopyrite was noted in ore from the West End vein a short distance above the 600-foot level (stope 600 C). Areas of pyrite adjacent to the chalcopyrite have not been replaced by covellite.

REPLACEMENT OF SPHALERITE AND GALENA BY COVELLITE OR ARGENTITE OR A MIXTURE OF THESE MINERALS.

In other parts of the specimen from the Shaft vein mentioned in the preceding section, covellite and mixtures of covellite and argentite were noted as formed by the replacement, mainly peripheral, of galena and sphalerite.

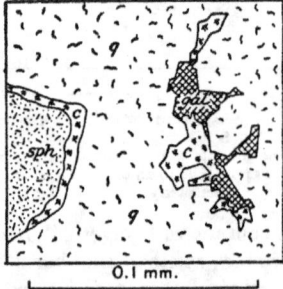

FIGURE 19.—Replacement of galena (*gal*) and sphalerite (*sph*) by covellite (*c*). *q*, Quartz. Camera lucida drawing of polished surface of ore from Shaft vein, 1,000-foot level of Tonopah Belmont mine.

These replacement deposits are shown in figures 19 and 20. In the supposed hypogene replacement deposits described on pages 19–25 sphalerite was rarely attacked, being in general stable under hypogene conditions.

In the specimen shown in figure 20 the width of the replacement rim does not change in passing from parts that are wholly argentite to parts that are wholly covellite. This feature was also noted in replacement borders of argentite and electrum developed around galena, as shown in Plate IX, *B*. In other parts of the same specimen galena is replaced by argentite, but there is a transition band of the unidentified galena-like mineral between the argentite and the galena.

[1] Anthon, E. F., Ueber die Anwendung der auf nassem Wege dargestellten Schwefelmetalle bei der chemischen Analyse: Jour. prakt. Chemie, vol. 10, p. 353, 1837. Posnjak, Eugene, Determination of cupric and cuprous sulphides in mixtures of one another: Am. Chem. Soc. Jour., vol. 36, pp. 2475–2479, 1914.

NATIVE SILVER.

In the specimen from the Shaft vein in which the replacement deposits described above were noted small areas of native silver

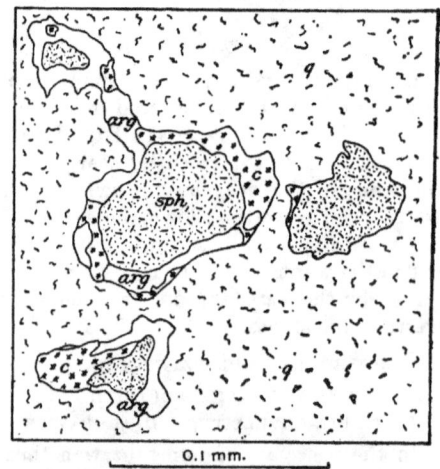

FIGURE 20.—Peripheral replacement of sphalerite (*sph*) by an association of covellite (*c*) and argentite (*arg*). *q*, Quartz. Camera lucida drawing of polished surface of ore from Shaft vein, 1,000-foot level of Tonopah Belmont mine.

are abundant. The silver-bearing parts of the specimen are commonly more or less porous and are apparently the most altered parts of the ore. The silver is usually associated with argenite and may, at least in part, replace it, as suggested by the relations shown in Plate XIII, *A*. In other parts of the same specimen, as shown in figure 21, silver and argentite

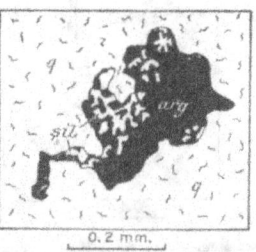

FIGURE 21.—Association of argentite (*arg*) and native silver (*sil*). Both minerals are probably a replacement of some hypogene sulphide. *q*, Quartz. Camera lucida drawing of polished surface of ore from Shaft vein, 1,000-foot level of Tonopah Belmont mine.

are so irregularly intergrown as to suggest contemporaneity.

As electrum is so common in the ores as a hypogene mineral, it would appear theoretically possible that native silver should occur also as a hypogene ore mineral. Close study of the ores, however, has failed to discover native silver in any situation where a supergene origin is not more probable than a hypogene origin. Furthermore, as gold was fairly abundant in the hypogene solutions, electrum would seem to be the more likely mineral to form.

RELATION BETWEEN SUPERGENE ENRICHMENT AND FRACTURING.

Deposition in open spaces and metasomatic replacement have both taken part in the supergene mineralization, but the former method of deposition has greatly predominated. Char-

rite, argentite, etc. The value of the ore decreased from the fault eastward and in the 100 feet west of winze 1,050 fell to $3 or $4 a ton. Low assays continued up to a northerly fault dipping 65° E. Just east of or above this fault high-grade ore averaging 0.3 ounce of gold and 30 ounces of silver to the ton was found for about 30 feet. Examination of this ore showed argentite along fractures in thin layers that could be readily stripped off and that are probably of supergene origin. There is half an inch or so of gouge along this fault, but the vein is not greatly displaced by it.

In the Extension No. 1 workings ore tenor in the Extension North vein appears to have

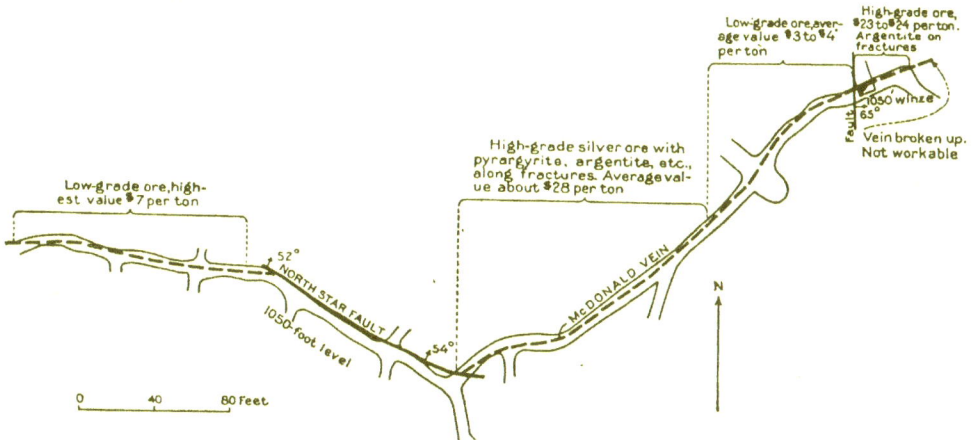

FIGURE 22.—Plan of a portion of the North Star mine workings, showing the relations between supergene enrichment and fault fractures.

acteristic deposits of argentite, pyrargyrite, polybasite, chalcopyrite, pyrite, native silver, barite, and carbonates on fractures cutting the hypogene ores are described above. The effect of such deposits on the average tenor of the ore in certain parts of a vein is in many places masked by the somewhat sporadic distribution of the hypogene replacement deposits; nevertheless a distinct relationship between late fracturing and ore tenor is recognizable in a number of places.

On the 1,050-foot level of the North Star mine (see fig. 22) the MacDonald vein west of the point where it was cut by the eastward-dipping North Star fault was not of exceptional grade; just east of or above this fault occurred the richest silver ore the mine has produced, carrying an abundance of pyrargy-

been influenced by the degree of fracturing near the Rainbow fault, which trends north-northeast and is not itself mineralized. The zone of maximum movement along this fault is marked by 2 feet of gouge dipping 50°–60° E. The North vein west of this gouge was little shattered, and its ore was not notably rich, but for 120 feet east of the gouge the vein was intensely shattered without serious displacement, was much oxidized, and was of high grade.

According to Mr. J. H. Dynant, of the Tonopah Extension Co., the ore of the McNamara vein on the 400-foot level was notably richer close to the Rainbow fault than away from it on either side. The fractured zone here is at least 20 feet wide, and the ore is highly brecciated and oxidized.

In the Tonopah Extension and West End, mines the most persistent and conspicuous fractures traversing the veins strike N. 10°–30° E. and are nearly vertical. Along most of these fractures the displacement has been slight, but some of the larger faults of similar trend and considerable displacement, such as the Rainbow fault, may belong to this series. Late deposits of argentite, polybasite, and carbonates are common along these fractures, which appear to have been influential in guiding the enriching solutions.

Just above the 500-foot level of the West End mine in stope 505 the West End vein, which strikes N. 80° W. and dips 35° N., is cut by a vertical fault striking N. 10° W. For 50 feet from the main fault the ore is traversed by a number of parallel subsidiary fractures. The ore of this 50-foot interval is said by Mr. J. W. Chandler to have been notably richer than that on either side. Examination showed that it carried argentite, chalcopyrite, barite, and ferruginous calcite on fractures.

In the Tonopah Belmont mine no such regular and extensive system of fractures traversing the ore was noted as in the western part of the camp, and it was reported that the few well-defined fractures that occur appear not to have influenced greatly the tenor of the ore. A probable exception to this statement is the Belmont fault, for some of the richest ore of the mine has been found in the Belmont vein for 50 feet east of (above) this fault on the 1,100 and 1,166 foot levels, where the ore was much fractured and oxidized. Irregular minor fractures are numerous in the ores of this mine, and it is probable that most of the ore was fairly permeable to descending solutions.

SUMMARY OF EVIDENCE OF SUPERGENE ORIGIN OF CERTAIN ORE MINERALS.

The supergene origin of the deep-yellow gold, silver haloids, hyaline silica, hydrous oxides of iron, oxides of manganese, malachite, calamine, dahllite, gypsum, and epsomite in the Tonopah ores is not open to queston.

The silver haloids are evidence of solution and redeposition of silver in the oxidized zone, and it is certainly to be expected that some of the silver taken into solution would be redeposited at greater depths as the sulphide or sulphosalts of silver. The notable enrich-

ment of the MacDonald vein above certain inclined faults (see p. 41) by deposits of pyrargyrite and argentite on fractures is concrete evidence that such deposition has taken place.

The thin coatings of crystalline quartz common on fractures may be of diverse origin and age. Though clearly formed after the main hypogene mineralization and after the ores had been fractured, some of this quartz was probably deposited by ascending solutions, as is shown by the presence of wolframite, an undoubted hypogene mineral, intergrown with such quartz in parts of the Belmont vein. Other quartz coatings are probably supergene, as for example those on oxidized ore below the Fraction dacite breccia. Not all the late fractures are coated with quartz, but where quartz is present it usually forms the first coating and the late silver minerals lie on the quartz, not intercrystallized with it.

Structural evidence of the supergene origin of some of the argentite, polybasite, and pyrargyrite in the ores has been presented, but the supergene origin of all occurrences of these minerals in late fractures and of the pyrite, chalcopyrite, barite, and carbonates in places associated with them can not be said to have been proved. Some of them may have been deposited by ascending solutions at a period distinctly later than the main hypogene mineralization, or by interaction between ascending and descending solutions. If they are hypogene they are probably later than the period of oxidation which preceded the deposition of the Fraction dacite breccia, for barite, calcite, and rarely pyrite coat highly oxidized ore just below this breccia. (See p. 32.) The writers believe that the probabilities favor the supergene origin of all the minerals except wolframite and a part of the quartz, which coat sharp-walled fractures cutting the typical hypogene ores.

Native silver appears to be absent from the deeper ores at Tonopah, even from those that carry argentite and sulphosalts of silver in abundance. It was noted only in ores that were somewhat oxidized, were more or less porous through solution of certain constituents, or showed other evidences of alteration, such as the presence of kaolin containing adsorbed soluble salts of copper. In places, its relations suggest that it has been formed by the alteration of argentite or polybasite. All the native silver at Tonopah is, therefore, believed to be supergene.

The replacement of chalcopyrite by bornite, of bornite by covellite and argentite, of sphalerite by covellite and argentite, and possibly of argentite by native silver, shown by a specimen from the Shaft vein, contrasts strongly with the replacement characteristic of most of the Tonopah ores. Under the conditions of the more usual and supposedly hypogene replacement chalcopyrite is stable (being itself a replacing mineral), argentite is never accompanied by native silver but on the contrary is in places accompanied by electrum, and sphalerite is only rarely replaced and then not by covellite and argentite but by chalcopyrite and polybasite. Obviously, these types of replacement occurred under physical and chemical conditions quite different from those under which most of the ores were formed. The loss of iron and the substitution of silver for copper involved in the replacement of chalcopyrite by bornite and of bornite by argentite, as well as the formation of native silver, are changes in harmony with well-known chemical tendencies in supergene enrichment, and to this process the writers attribute them.

NATURE OF SOLUTIONS PRODUCING SUPERGENE MINERALIZATION.

Where active oxidation of pyrite and chalcopyrite was in progress, acidity must, at least locally and transiently, have been imparted to the waters in contact with the ore. The abundance of primary carbonates (of one or more of the elements calcium, magnesium, iron, and manganese) in most of the Tonopah ores, however, favors the neutralization of acidity early in the descent of the solution. This inferential conclusion appears to tally with the observed facts, for, as shown in the detailed descriptions that precede, the argentite and polybasite in late fractures are in many places intergrown with or even wholly inclosed in calcite or related carbonates. Such mineral associations were evidently deposited by neutral or possibly alkaline and not by acid solutions. If such deposits in fractures have been correctly interpreted as supergene, it follows that the supergene solutions were not acid at the places at which these minerals were deposited, whatever may have been their reaction earlier in their descent. Even the silver haloids were, locally at least, deposited by neutral solutions, as shown by their intercrystallization with malachite and calcite. Barite is associated with supposedly supergene argentite and polybasite in much the same manner as the carbonates. The mineralogic evidence indicates therefore (1) that the solutions which produced supergene enrichment became neutral early in their downward journey; (2) that they carried carbonates and sulphates in abundance, with lesser amounts of halogens and phosphates; and (3) that they transported silver, copper, iron, manganese, arsenic, antimony, and possibly other metals.

In the hope of obtaining further evidence concerning the character of the solutions that produced the supergene enrichment, a sample of descending mine water was collected in the West End mine and was analyzed by Chase Palmer. (See table on p. 29). This water flowed down through the broken ore and waste filling of a large caved stope between the 400 and 500 foot levels and issued in a small stream at raise 524 on the 500-foot level. In the stope it is in contact with ore and with the air, and the conditions simulate those originally existing a short distance below the natural outcrop of the veins. The water as tested in the mine was decidedly acid to methyl orange and had eaten holes where it dripped on corrugated iron at the foot of raise 524.

The analysis shows that this is essentially a sodium and calcium sulphate water, carrying also manganese and chlorine radicles in abundance. Its concentration is nearly twice that of the Mizpah deep water and nearly five times that of the Belmont deep water. The general relations between its components, computed in accordance with the Palmer method, are as follows:

Alkali radicles balanced by strong acid radicles....	26.8
Alkali earth radicles balanced by strong acid radicles...	57.8
Metallic radicles balanced by strong acid radicles..	14.2
Free strong acid (?)............................	1.2
	100.0

The slight excess of strong acid radicles in the analysis is within the permissible limits of experimental error, and it is therefore uncertain whether the observed acidity of the water toward methyl orange is due to the presence of free acid (SO_4 in balance with H) or to the hydrolysis and ionization of the small amounts of salts of strong acids and weak bases present.

ferrous and aluminous sulphate in aqueous solution both giving acid reactions. The entire absence of carbonate and bicarbonate radicles, in spite of the fact that the water has percolated through carbonate-bearing ore, is a natural consequence of the acidity of the water, as a result of which carbonates would be converted into sulphates with liberation of carbon dioxide.

It appears that this water may present some analogy to the waters that accomplished the supergene enrichment in the early part of their descent, before neutralization and before they had acquired their full load of dissolved materials. It is apparently not an analogue of the supergene solutions at the stage when they deposited most of the supergene silver minerals, for at that stage they appear to have been neutral and were depositing also carbonates.

PROBABLE PERSISTENCE OF ORES IN DEPTH.

The deepest developments in the Tonopah mines have shown that in certain veins the degree of mineralization gradually decreases with increasing depth. The deep ores of such veins carry more quartz and less sulphides than the ores higher up; microscopic study shows, however, that the sulphides are the same species as in the ores above and that replacement deposits of the type described as probably hypogene are present, though naturally not in such abundance as in the ores richer in sulphides. The Belmont vein on the 1,500-foot level of the Tonopah Belmont mine, the combined vein on the 770-foot level of the Extension No. 1 workings, and the Favorite vein on the 1,200-foot level of the Tonopah Belmont mine afford good examples of such diminution in value with depth.

The Belmont vein on the 1,500-foot level shows sulphides only in a few places; much of it is brecciated wall rock traversed by veinlets of white or dark-gray quartz, and no ore in commercial quantity has yet been found. Ore from a point 40 feet east of raise No. 2, when examined microscopically, was seen to carry scattered grains and small aggregates of galena, sphalerite, chalcopyrite, pyrite, and polybasite in a quartz gangue. The polybasite, at least in part, replaces galena, and there are narrow transition rims of probable lead-silver sulphide between. A sample of this ore assayed by the Bureau of Mines showed 0.08 ounce of gold and 3.90 ounces of silver to the ton. The hand specimen shows very slight oxidation along fractures, and kaolin occurs as a late deposit on the quartz crystals in some vugs.

Ore from the 770-foot level of the Extension No. 1 workings is highly quartzose and carries scattered minute grains of pyrite and, rarely, small dark-colored aggregates of other sulphides. A polished surface of a piece comparatively rich in dark sulphides shows galena, sphalerite, pyrite, chalcopyrite, and polybasite. The polybasite is peripherally replaced by an aggregate of argentite and chalcopyrite. The hand specimens show hydrous iron oxides along some fractures and manganese oxides along others.

In contrast to the veins just mentioned, other veins show strong mineralization on the deepest levels to which they have been developed. Ore from the Murray vein, for example, on the 1,260-foot level, is rich in sulphides and is equal in grade to much of the ore higher up. This ore is unoxidized. Microscopic study of a polished specimen shows the presence of galena, sphalerite, pyrite, and chalcopyrite, in addition to quartz and a pink manganiferous carbonate. In places the galena has been peripherally replaced by argentite and electrum, as shown in Plate IX, B. Some kaolin has been deposited in small vugs.

The decrease in tenor with increasing depth in the veins of the Tonopah Belmont mine is concomitant with the passage of the veins from the Mizpah trachyte into the West End rhyolite and is doubtless dependent, at least in part, upon the differing susceptibilities of the two rocks to fracturing and replacement. The mineralization, though varying in degree, was continuous from one formation to the other.

Although obviously all veins must terminate in depth as well as laterally, the evidence from microscopic studies that many of the silver minerals at Tonopah are hypogene (primary) and that they occur in ore from the deepest workings offers substantial encouragement to deep mining. In some veins change in wall rock was probably a more important factor than mere increase in depth in causing the decreased ore tenor.

OXIDATION BANDING IN WALL ROCKS.

A feature of no economic importance but of some scientific interest is the regularly banded distribution of hydrous oxides of iron in certain of the wall rocks. This feature is especially well shown in material from the dump of the Valley View shaft, and most of the specimens studied were from that source. Similar banding was noted in place in the walls of the South vein, Desert Queen mine, at a depth of about 500 feet.

The agents producing this banding evidently gained access to the wall rock through fractures and from them permeated the rock. Polygonal blocks bounded by fractures show completely concentric structures such as are illustrated in Plate XIII, B.

The iron oxides appear to have been derived from the oxidation of pyrite, for in many specimens cores carrying abundant disseminated grains of pyrite remain. This is true of the specimen illustrated in Plate XIV, A, the light-colored, unbanded portion carrying abundant pyrite grains. Plate XV shows photomicrographs taken at the junction of the oxidized and unoxidized areas. Plate XV, A, taken in ordinary light, shows the scattered crystals of pyrite (black) in the unoxidized portion and irregular blotches of iron oxide, probably amorphous, in the oxidized portion. In the oxidized portion are visible several voids corresponding in size and shape to the pyrite crystals and evidently representing their former presence. Plate XV, B, gives the appearance of the same field in polarized light and shows that the oxidation banding is independent of the original texture of the rock.

Quartz veinlets that traverse the rock were at first believed to be later than the banding, but closer study leads to the conclusion that they were earlier. In places they have had surprisingly little influence on the course of the oxidation bands. Near some very late fractures the iron oxides have been leached.

A very interesting example of oxidation banding in fine-grained felsite from the vicinity of the North Star shaft is shown in Plate XIV, B. The senior writer was at first inclined to interpret this as a "development" by oxidation processes of a banded distribution of pyrite or ferruginous carbonates produced during hydrothermal metamorphism; microscopic study shows, however, that the position of the bands of iron oxide is wholly independent of the distribution of the pyrite from which it was plainly derived.

It has been suggested by A. C. Spencer that the banding is the result of alternate wetting and drying during many seasons. This explanation has much to recommend it in a region like Tonopah, where cyclonic storms are mainly confined to the winter months, the remainder of the year being rainless except for occasional showers. Furthermore, occasional drying permitting access of air would appear to be necessary for the oxidation of the pyrite.

The possibility of the development of such oxidation banding by diffusion processes analogous to the formation of the so-called Liesegang rings has been pointed out by Liesegang[1] and unquestionably merits serious consideration, although the progress of such processes must necessarily have been modified by alternations of wet and dry periods. In this connection two experiments carried out by the writers may be of interest.

In the first experiment the central portion of a straight glass tube about half an inch in internal diameter was filled with a solution of agar agar in water. This was retained in place by a wad of cotton until it hardened. After the agar agar had set to a firm jelly a solution of N/20 ferric sulphate was placed in one end of the tube and a solution of N/16 sodium hydroxide in the other end; the ends were then corked and sealed. After two days a faint brown band of ferric hydroxide had developed at the point marked A in Plate XVI, A. At the end of another two days the band A had not moved but was better defined and a new band of ferrous hydroxide (green) had developed at the point marked B. By the successive formation of new bands shown, from left to right in Plate XVI, A (from $Fe_2(SO_4)_3$ toward NaOH), the tube at the end of 17 days presented the appearance shown in the illustration. Bands once formed did not shift appreciably. The agar agar appears to have exerted a reducing influence on the ferric sulphate, so that most of the rings were of ferrous hydroxide instead of ferric hydroxide as was expected.

In a second experiment the agar agar was placed in the lower part of a U tube; sodium

[1] Liesegang, R. E., Geologische Diffusionen, pp. 106–116, 1913.

hydroxide was placed above it in one arm and ferrous hydroxide in the other. The appearance at the end of one month is shown in Plate XVI, *B*. Most of the rings were green, but those last formed, at the point marked *A*, were brown, evidently because of oxidation of the iron-bearing solution by the air.

Rings were also produced in compacted filter paper under conditions similar to those of the first experiment.

ANALYSES OF ORE, BULLION, ETC.

The following analyses, published by Mr. A. H. Jones[a] and inserted here with his permission, have significance in indicating the presence of certain elements, notably selenium, whose presence is not apparent from a mineralogic examination, and in showing that certain others, such as nickel and cobalt, which are characteristic of some silver deposits, are probably absent. For comparison the analysis by Hillebrand published in Spurr's report[b] is included.

Analyses of Tonopah ores, concentrates, bullion, etc.

	1	2	3	4	5
SiO_2	72.00	15.18	......	5.50	......
Al_2O_3	.42		0.42	2.00	
CaO	3.10	3.70	.80	.10	
MgO	2.95	1.49	......	Trace.	......
$CaCO_3$				3.60	
CO_2		6.34			
NaCl and KCl				1.00	
$FeSO_4$				.26	
Fe	2.10	9.87	29.80		
Mn	1.60	1.36	1.10		
Cu	1.09	1.32	.60	2.10	1.02
Ni and Co	None.				
Pb	1.50	6.21	1.30	2.36	2.41
Zn	3.00	5.84	.60	4.70	.06
Cd	None.			1.00	
As	Trace.	.19			
Sb	.10	.92			
Bi	None.				
S	2.60	Not det	31.60		.17
Se	.20	2.56		1.40	1.80
Te	Trace.	None.			
Au	.06	.82	} 1.60	74.23	93.23
Ag	4.88	25.92			
Undetermined	c4.40		1.58	d1.75	b1.31
Insoluble			30.60		
	100.00	81.72	100.00	100.00	100.00

[a] By difference, mainly Na, K, and combined H_2O.
[b] By difference.
[c] Jones, A. H. (superintendent of mills, Tonopah Belmont Development Co.), The Tonopah plant of the Belmont Milling Co.: Am. Inst. Min. Eng. Trans., vol. 52, pp. 1731–1738, 1916.
[d] Spurr, J. E., Geology of the Tonopah mining district, Nev.: U. S. Geol. Survey Prof. Paper 42, p. 89, 1905.

1. Analysis by A. H. Jones of piece of rich silver ore from Belmont mine; exact location in mine unknown.
2. Analysis by W. F. Hillebrand, in the laboratory of the United States Geological Survey, of sulphides separated by crushing and panning from sample of rich sulphide ore from Montana vein at depths of 460 to 512 feet in Montana Tonopah mine.
3. Partial analysis by A. H. Jones of concentrates from Belmont mill.
4. Analysis by A. H. Jones of zinc box precipitate, Belmont mill.
5. Analysis by A. H. Jones of bullion, Belmont mill.

The cadmium shown in analysis 4 very possibly represents a concentration of cadmium occurring as an impurity in the zinc used for precipitation. It is noteworthy that the ratio of gold to selenium is very nearly the same in both analyses 1 and 2.

IDENTIFICATION OF METALLIC MINERALS IN POLISHED SECTIONS.

Some of the methods of identification of minerals in polished section that were used in these studies may include a few new points of possible service to other students of ore deposits.

Whenever possible small fragments of the mineral to be identified were chipped from the edges of the polished specimen with a stout needle or very sharp steel prod and tested before the blowpipe or in a wet way.

Needles of several sizes mounted in the ends of wooden penholders were found very useful as prods for testing the hardness, streak, sectility, etc., of minerals.

Most reagents for tarnishing the minerals can be conveniently applied on a strip of blotting paper about half an inch wide and 3 inches long. One end of this strip is moistened with the reagent and is rubbed lightly back and forth over the polished surface of the mineral, the progress of the tarnishing being watched under the low or moderate powers of the microscope and checked at the desired stage by simply rubbing with the dry end of the blotting paper.

Reactions aiding in the identification of certain minerals are the following:

The bright silvery luster of galena in polished sections, its lead-gray streak, and the development of triangular pits due to the tearing out of inverted pyramids of galena between the three sets of cleavage planes during polishing (see, for example, Pl. VII, *B*) have usually been considered sufficient to identify it. The writers discovered, however, that another mineral, probably a lead-silver sulphide, also possesses these characteristics. It was found that the two could be differentiated by treatment with ordinary commercial hydrogen peroxide (H_2O_2). This reagent tarnishes galena brown but does not affect the supposed lead-silver sulphide or any of the other silver minerals with which galena is commonly associated. It is therefore a reagent of great utility in studying the relations between galena and associated silver minerals. The tarnish it produces on galena appears slowly, can be quickly checked, and can be easily removed by a little polishing with rouge on a bit of cotton.

The polished surfaces of argentite usually appear slightly "rougher" under the microscope than most of the other sulphides with which it is associated except pyrite; this feature is well shown in Plate III, *B*. Pricking with a fine needle reveals its soft sectile character. It is unaffected by hydrogen peroxide and silver nitrate but is rapidly tarnished brown by saturated mercuric chloride solution.

Polybasite is brittle and gives a black streak. Its polished surfaces are unaffected by hydrogen peroxide but are tarnished reddish brown by silver nitrate solution and rapidly tarnished brown by saturated mercuric chloride solution.

Pyrargyrite is brittle and gives a red streak. It is unaffected by hydrogen peroxide but is tarnished brown by silver nitrate solution, though much more slowly than polybasite. Short treatment with saturated mercuric chloride solution tarnishes it pale brownish yellow; polybasite by the same treatment is tarnished a pronounced brown. The polybasite shown in Plate IV, *B*, has been differentiated from the pyrargyrite which incloses it by such treatment. Before treatment the two minerals were hardly distinguishable from each other.

Silver nitrate is a useful reagent for tarnishing chalcopyrite, which becomes reddish yellow, then deep amber, and finally madder-red; argentite, galena, and sphalerite are unaffected by the same treatment.

SUMMARY OF MORE IMPORTANT CONCLUSIONS.

1. The hypogene or primary ores have been modified in places by oxidation and enrichment through the agency of the air and oxygenated solutions originating at or near the surface. The high silver content of much of the ore obtained in the past and of some ore now remaining is unquestionably due in part to these processes.

2. There is evidence not only of recent oxidation of the ores but also of at least one period of ancient oxidation, and supergene sulphide enrichment was probably an accompaniment of each of these periods.

3. The rich silver ores now being mined at Tonopah are probably in the main of hypogene or primary origin.

4. Mining has shown that in certain veins the primary sulphides become less abundant with increasing depth, though the same species are present; mere increase in depth may account for this change in some veins, for every vein must finally end in depth as well as laterally; in many veins change in wall rock has been at least a contributing factor. The veins developed by other deep workings are heavily mineralized and of high grade, and the geologic evidence is favorable to the persistence of rich primary silver ores to depths considerably greater than those yet attained in the mining operations.

Although hot ascending waters are encountered in a number of the deeper workings, there is little evidence that these waters are now depositing ores.

REFERENCES.

The following list gives the more important publications on the Tonopah district:

BURGESS, J. A., The geology of the producing part of the Tonopah mining district: Econ. Geology, vol. 4, pp. 681-712, 1909.

—— The halogen salts of silver and associated minerals at Tonopah, Nev.: Econ. Geology, vol. 6, pp. 18-21, 1911.

EAKLE, A. S., The minerals of Tonopah, Nev.: California Univ. Dept. Geology Bull., vol. 7, pp. 1-20, 1912.

LOCKE, AUGUSTUS, The geology of the Tonopah mining district: Am. Inst. Min. Eng. Trans., vol. 43, pp. 157-166, 1912.

ROGERS, A. F., Dahllite (podolite) from Tonopah, Nev.: Am. Jour. Sci., 4th ser., vol. 33, pp. 475-482, 1912.

SPURR, J. E., Geology of the Tonopah mining district, Nev.: U. S. Geol. Survey Prof. Paper 42, 1905.

—— Report on the geology of the property of the Montana Tonopah Mining Co., Tonopah, Nev., published by the company, 1910. Abstract in Min. and Sci. Press, vol. 102. pp. 560-562, 1911.

—— Geology and ore deposits at Tonopah, Nev.: Econ. Geology, vol. 10, pp. 713-769, 1915.

INDEX.

A. BANDED HYPOGENE ORE.

Enlarged 1½ diameters. The light-gray areas are mainly a fine intergrowth of ferruginous rhodochrosite and quartz;
the dark-gray to black areas are mainly quartz carrying hypogene sulphides in numerous very minute grains.
The radial structure of some of the rhodochrosite-quartz intergrowths at right angles to the banding is dimly
shown. Belmont vein, 1,000-foot level.

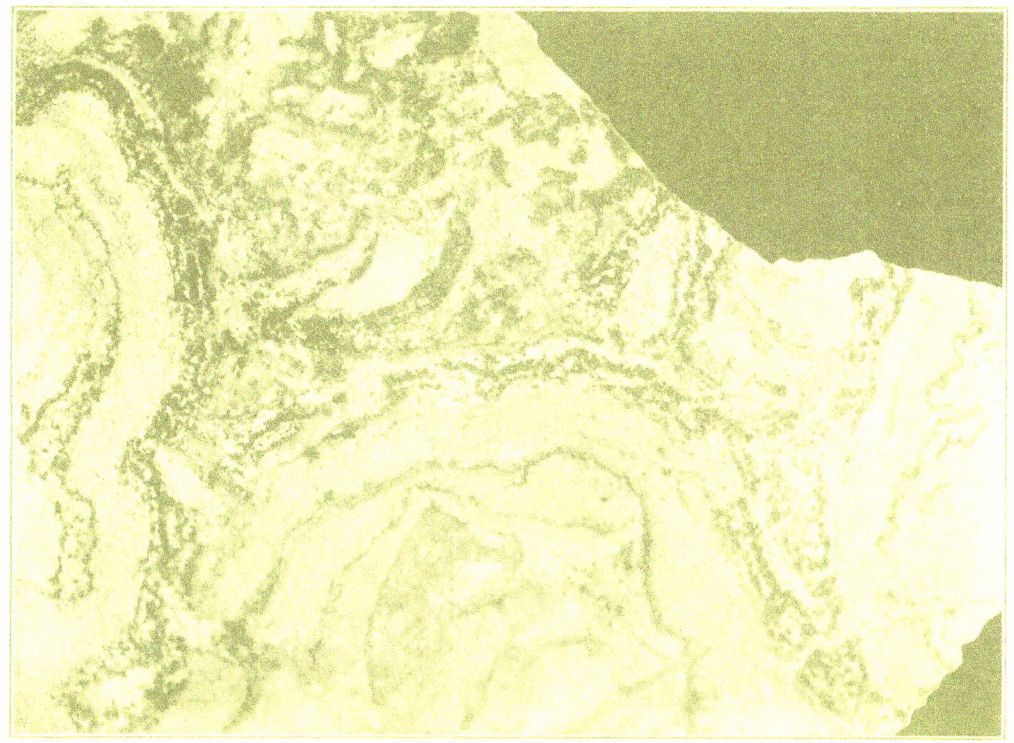

B. BANDED HYPOGENE ORE.

Murray vein, Tonopah Extension mine, just below 950-foot level. One-half natural size.

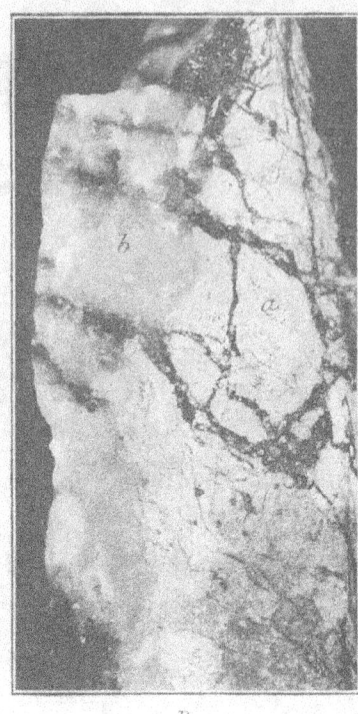

A.

B.

A, B. SUPPOSED HYPOGENE VEINLETS OF PYRARGYRITE.

The veinlets pass from areas of partly altered wall rock (*a*) into quartz areas (*b*) and become less well defined in the quartz. Tonopah Extension mine, exact location uncertain. Natural size.

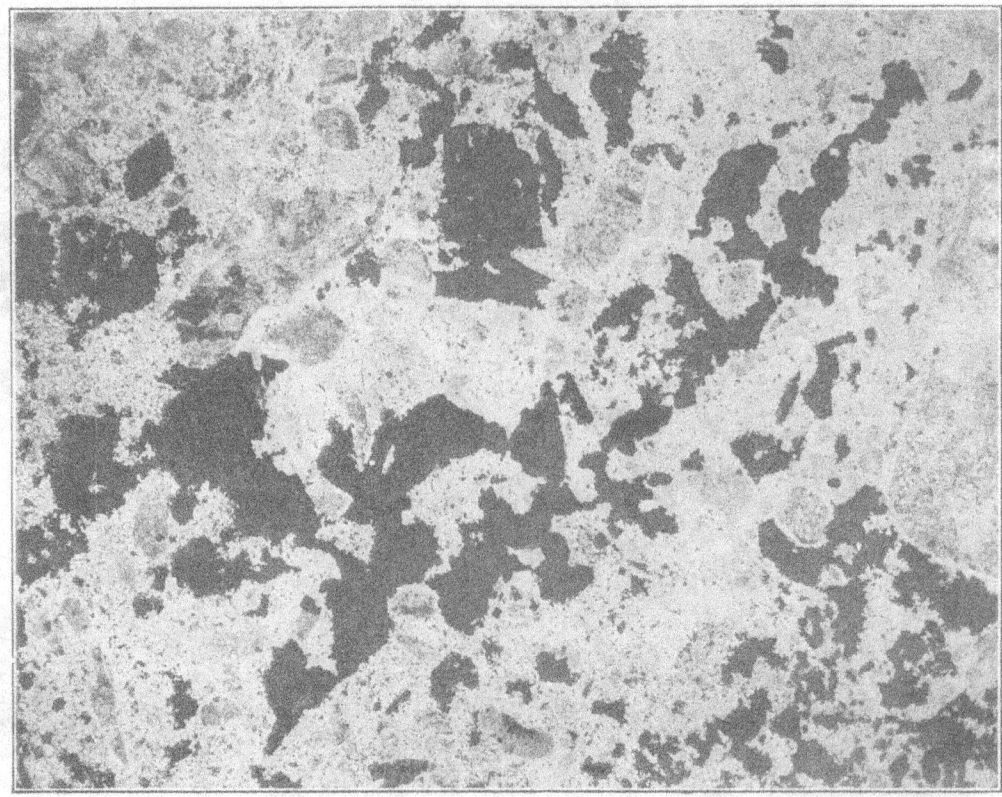

C. BRECCIA OF QUARTZ AND SERICITIZED WALL ROCK PARTLY REPLACED BY PYRARGYRITE (BLACK).

Photomicrograph of polished surface of ore. Tonopah Extension mine, exact location uncertain. Enlarged 25 diameters.

A. HYPOGENE (?) INTERGROWTH OF QUARTZ (DARK) AND PYRARGYRITE (LIGHT).

Photomicrograph of polished surface of ore from vein between 400 and 500 foot levels, Last Chance mine. These minerals are believed to be contemporary results of rock replacement.

B. IRREGULAR VEINLET OF SPHALERITE (SMOOTH LIGHT GRAY) AND ARGENTITE (ROUGH LIGHT GRAY), BORDERED BY QUARTZ (DARK GRAY), WITH SCATTERED IRREGULAR AREAS OF SPHALERITE.

The sphalerite-argentite veinlet is believed to mark a line of easy replacement in the original rock. All the minerals are interpreted as alpha hypogene. Photomicrograph of polished surface of ore from stope 600 C, West End vein, West End mine.

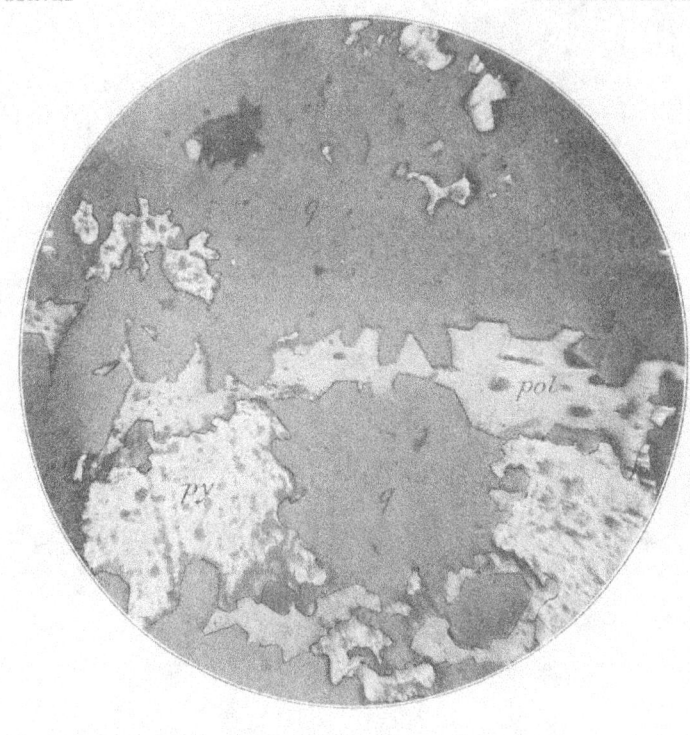

1 mm

A. IRREGULAR INTERGROWTH OF POLYBASITE (*pol*), PROBABLY ALPHA HYPOGENE, WITH QUARTZ (*q*) AND PYRITE (*py*).

Photomicrograph of polished surface of ore just above 500-foot level. West End vein, West End mine.

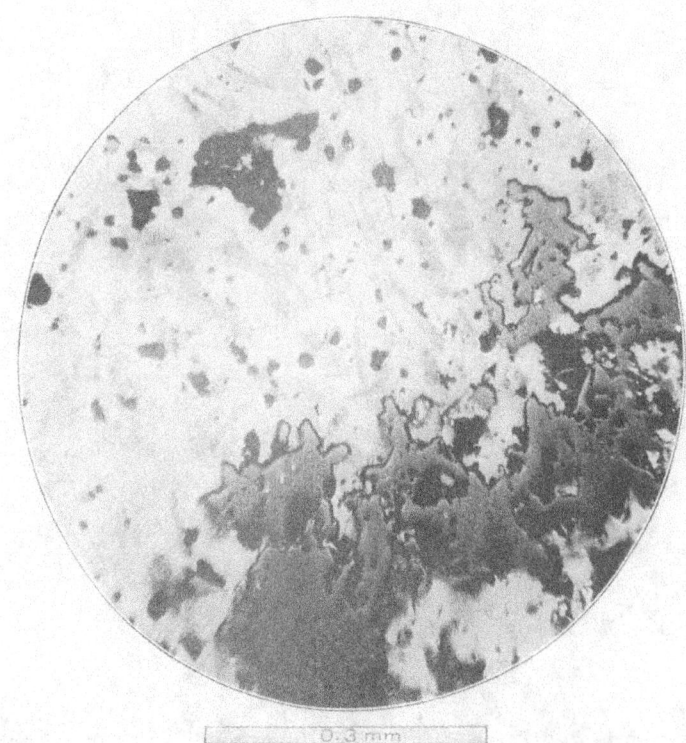

0.3 mm

B. POLYBASITE (LIGHT GRAY) WITH BLADELIKE OUTLINES, INCLOSED BY PYRARGYRITE (WHITE).

The dark gray is quartz. The polybasite has been made visible by brief treatment with mercuric chloride solution. Photomicrograph of polished surface of ore, Tonopah Extension mine, exact location uncertain.

A. ELECTRUM (STIPPLED) IRREGULARLY INTERGROWN WITH SPHALERITE AND QUARTZ.

The argentite (*arg*) shown in this photograph was probably formed by the replacement (beta hypogene) of galena (*gal*); all the other minerals, including the electrum, are believed to be alpha hypogene. Photomicrograph of polished surface of ore from 1,050-foot level, Macdonald vein, North Star mine. *sph,* Sphalerite; *q,* quartz.

B. ELECTRUM SHOWING CRYSTAL OUTLINES ASSOCIATED WITH QUARTZ AND SULPHIDES.

The argentite was very probably formed by the replacement of galena, but all the other minerals are believed to be alpha hypogene. Photomicrograph of polished surface of ore from 1,200-foot level, Murray vein, Tonopah Extension mine. *chal,* Chalcopyrite; *sph,* sphalerite; *arg,* argentite; *elec,* electrum; *q,* quartz.

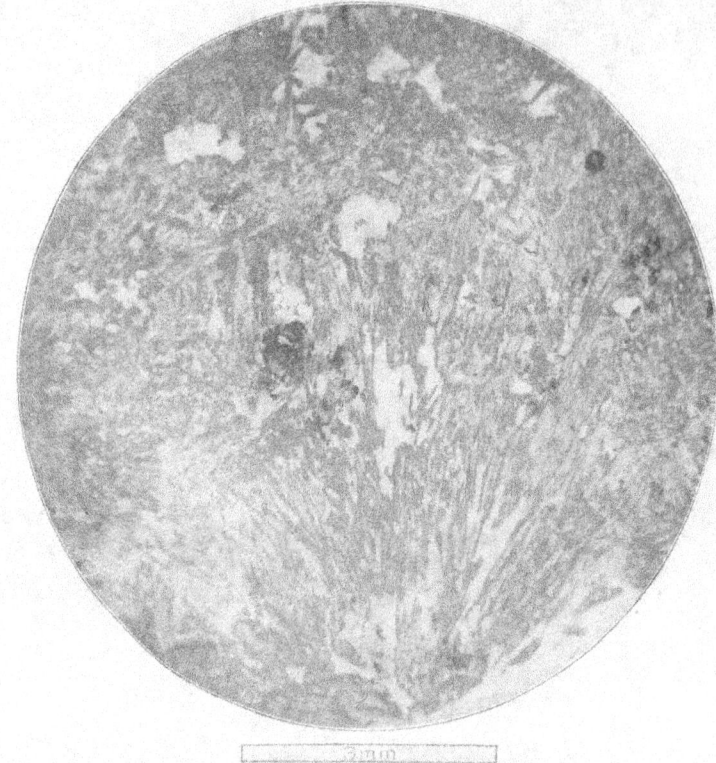

A. RADIATING INTERGROWTHS OF QUARTZ (WHITE) AND FERRUGINOUS RHODOCHROSITE
(GRAY) WITH SMALL PATCHES OF FINELY DIVIDED SULPHIDES (BLACK).

Photomicrograph (ordinary light) of thin section of ore from 1,000-foot level, Belmont vein, Tonopah-Belmont mine.

B. IRREGULAR FINE AGGREGATES OF CARBONATE, ARGENTITE, AND CHALCOPYRITE (*chal*)
INTERGROWN WITH ORE MINERALS IN LARGER GRAINS.

Photomicrograph of polished surface of ore from Favorite vein, Tonopah-Belmont mine. *sph*, Sphalerite; *q*, quartz.

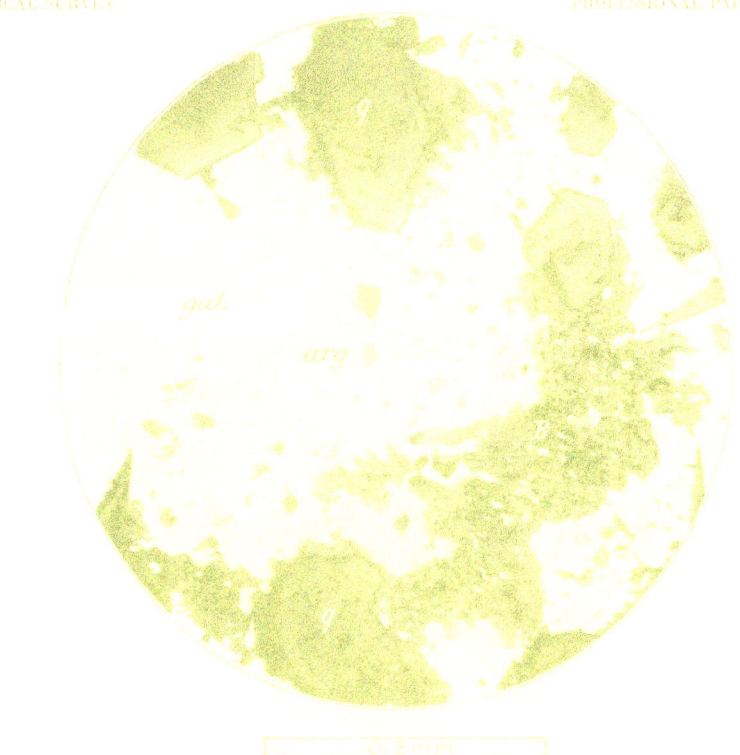

A. TRANSITION FROM AREAS WHOLLY GALENA (*gal*) INTO AREAS (*J*) THAT ARE MAINLY GALENA WITH SOME CARBONATE (ROUGH DARK GRAY) AND ARGENTITE (*arg*) AND FINALLY INTO AREAS (*b*) THAT ARE MAINLY CARBONATE WITH SOME ARGENTITE, GALENA, CHALCOPYRITE, AND ELECTRUM, NAMED IN THE ORDER OF ABUNDANCE.

Photomicrograph of polished surface of ore from 1,000-foot level, Favorite vein, Tonopah-Belmont mine.   *q*, Quartz.

B. REPLACEMENT OF GALENA (*gal*) BY POLYBASITE (*pol*).

Between these two minerals occur narrow bands of an unidentified transition mineral (not visible in the picture), and the very narrow light-colored bands occur (*J*) that mark the initiation of the replacement processes along the contacts between galena crystals consist of this mineral.   Photomicrograph of polished surface of ore from Favorite vein, Tonopah-Belmont mine, depth uncertain.

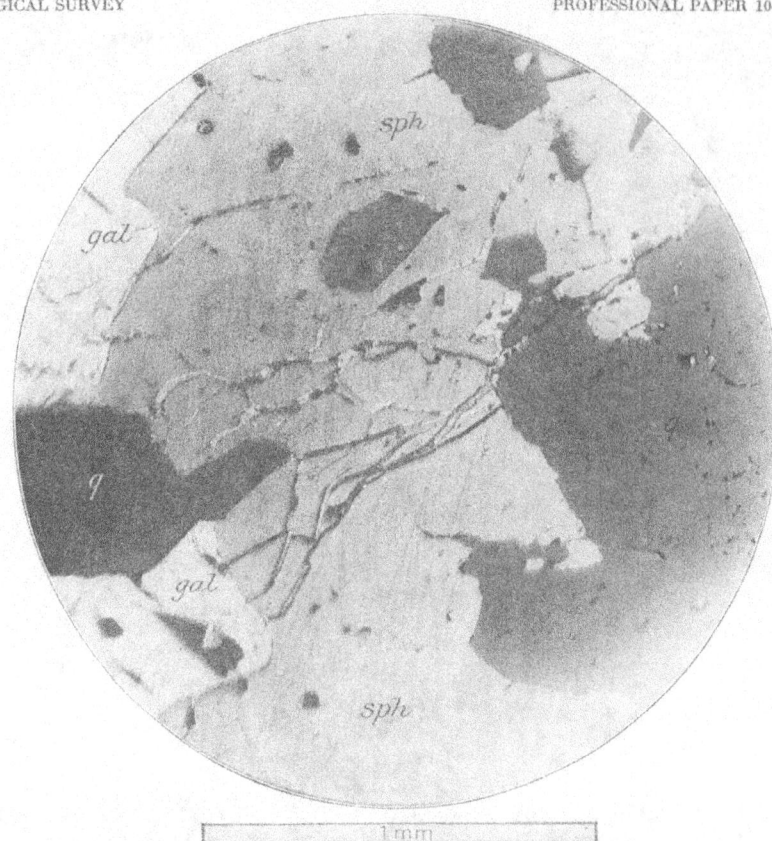

A. SPHALERITE (*sph*) TRAVERSED BY MINUTE VEINLETS OF CHALCOPYRITE AND POLYBASITE.

Some of the veinlets in sphalerite are continuous with areas of the same minerals replacing galena (*gal*); this feature
   does not show in this picture but is illustrated by figure 7, drawn from the same specimen. The veinlets are
   in the main formed by replacement, though possibly in small part fracture fillings. Photomicrograph of
   polished surface of ore from hanging-wall branch of Belmont vein, 1,200-foot level, Tonopah-Belmont mine.
   q, Quartz.

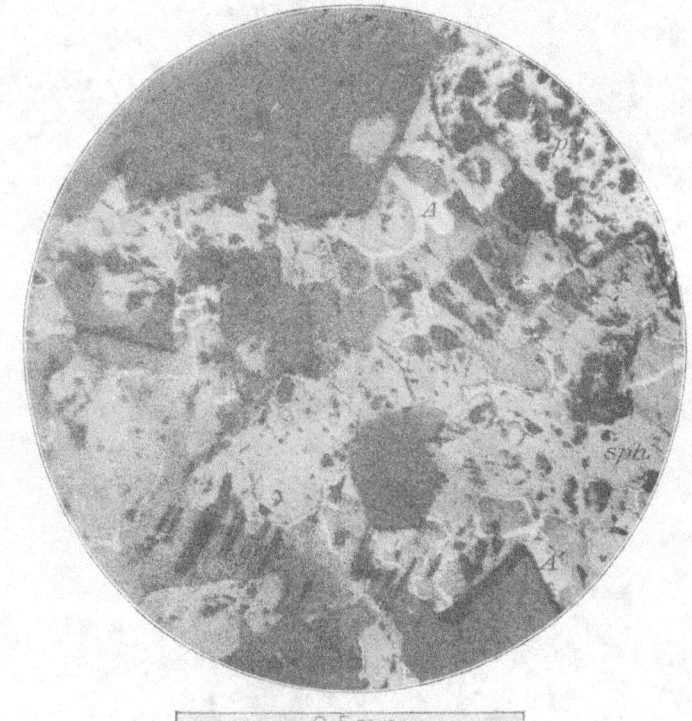

B. GALENA (*gal*) BORDERED BY AN UNDETERMINED REPLACEMENT MINERAL, PROBABLY A
LEAD-BEARING SILVER SULPHIDE.

In a few places, as at *A* and *A'*, the replacement process has proceeded further, with the development of argentite.
   Photomicrograph of polished surface of ore from Favorite vein, Tonopah-Belmont mine, depth not known.
   q, Quartz; *py*, pyrite; *sph*, sphalerite.

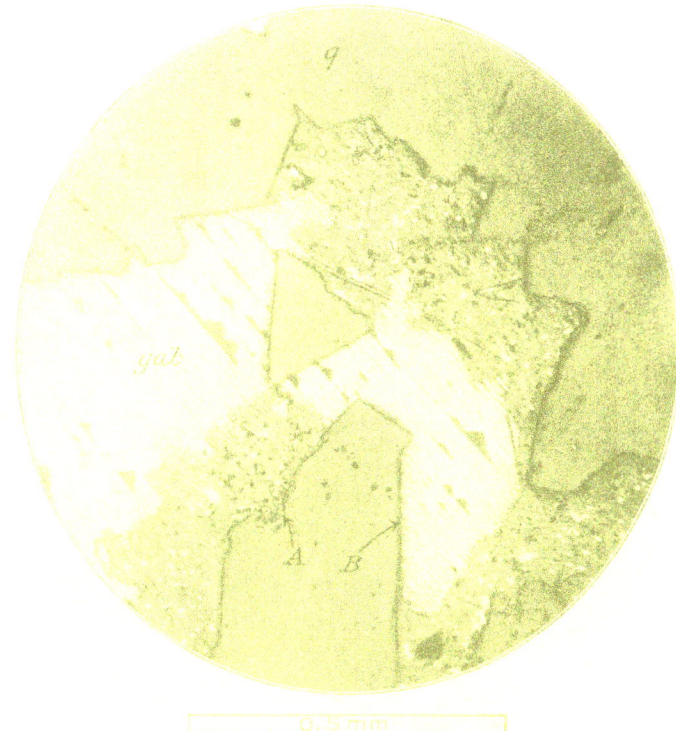

A. PARTIAL REPLACEMENT OF GALENA (*gal*) BY A FINE AGGREGATE OF A LIGHT-COLORED
CARBONATE, ARGENTITE, AND SOME CHALCOPYRITE.

There has also been a slight replacement of the quartz (compare the ragged boundary of the quartz (*q*) at *A* with
the smooth boundary at *B*). Photomicrograph of polished surface of ore from Favorite vein, Tonopah-Belmont
mine, depth uncertain.

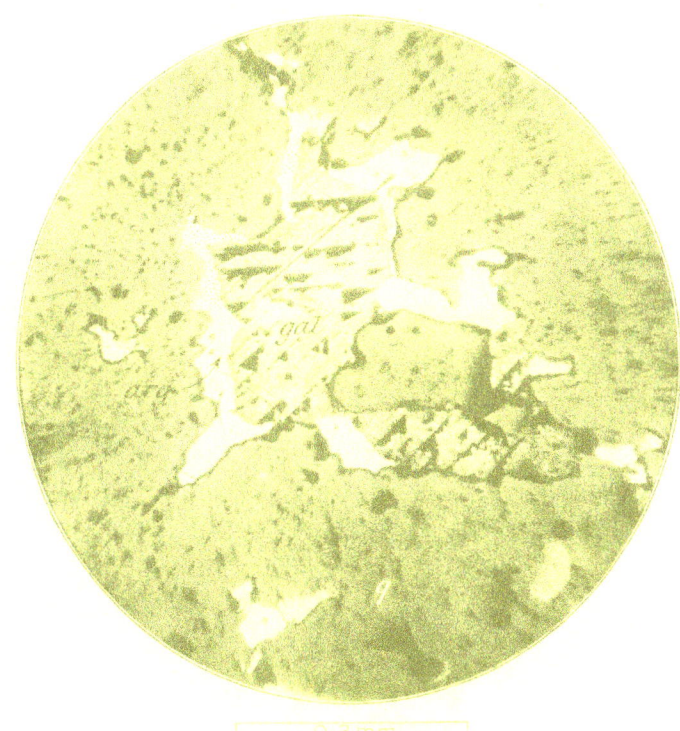

B. PERIPHERAL REPLACEMENT OF GALENA (*gal*) BY AN ASSOCIATION OF ARGENTITE (*arg*) AND
ELECTRUM (STIPPLED).

Note the smoothness of the galena border in passing from contact with argentite to contact with electrum. There
has also been replacement of the type shown in figure 6 along the contact between two galena crystals of
different orientations. Photomicrograph of polished surface of ore from 1,250-foot level, Murray vein, Tonopah
Extension mine. *q*, Quartz.

1 mm

A. PHOTOMICROGRAPH SHOWING TUFFACEOUS CHARACTER OF FRACTION DACITE BRECCIA.

Near station 1147 on 300-foot level of West End mine.

B. FERRUGINOUS CALCITE COATING HIGHLY OXIDIZED ORE JUST BELOW FRACTION DACITE
BRECCIA.

From stope 484 not far above 400-foot level, West End vein, West End mine. Two-thirds natural size.

A. BRANCHING DEPOSIT OF ARGENTITE, PROBABLY SUPERGENE, ON WALL OF FRACTURE
IN QUARTZOSE HYPOGENE ORE.

Small amounts of finely crystalline chalcopyrite are associated with the argentite. Specimen from 1,130-foot level,
Macdonald vein, North Star mine, in collection of A. E. Lowe, Tonopah. Natural size.

B. LICHEN-SHAPED DEPOSITS OF ARGENTITE, PROBABLY SUPERGENE, WITH A LITTLE POLY-
BASITE, ON WALL OF FRACTURE IN HYPOGENE RHODOCHROSITE-RICH ORE.

Not far below 900-foot level, Murray vein, Tonopah Extension mine. Natural size.

A. POLYBASITE, PROBABLY SUPERGENE, PARTLY FILLING FRACTURES IN WHITE, OPAQUE QUARTZ.

In a few places finely crystalline chalcopyrite and in one place a little native silver are deposited on the polybasite. Macdonald vein, Montana-Tonopah mine.

B. BRANCHING DEPOSITS OF PYRARGYRITE AND POLYBASITE, PROBABLY SUPERGENE, ON WALLS OF FRACTURE IN WHITE QUARTZ.

Montana-Tonopah mine, exact location uncertain. About natural size.

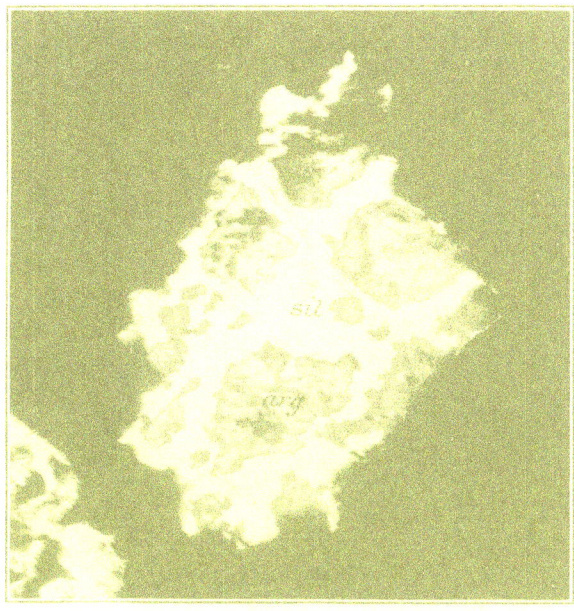

A. ASSOCIATION OF SILVER (*sil*) AND ARGENTITE (*arg*) SUGGESTING REPLACEMENT OF THE
SULPHIDE BY THE NATIVE METAL.

Photomicrograph of polished surface of ore from 1,000-foot level, Shaft vein, Tonopah-Belmont mine. The argentite boundaries have been outlined to render them more distinct. The knob of polished silver and argentite is surrounded by soft porous material which shows black in the photograph.

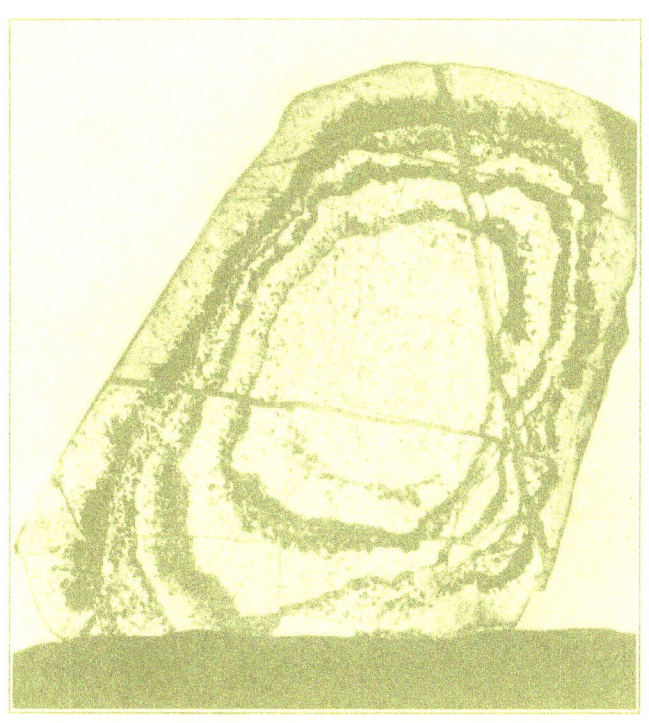

B. OXIDATION BANDING IN MIZPAH TRACHYTE.
From walls of Valley View vein. Natural size.

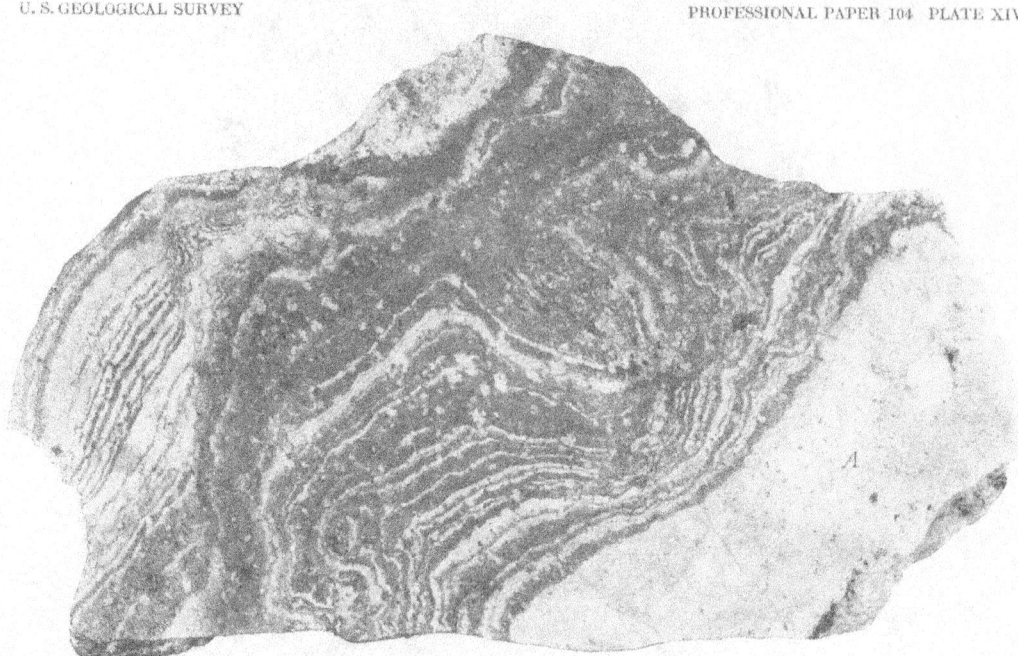

A. OXIDATION BANDING IN MIZPAH TRACHYTE ADJACENT TO VALLEY VIEW VEIN.

About two-thirds natural size. Small crystals of pyrite are abundant in the unoxidized portion (*A*) but are absent from the banded portion of the vein.

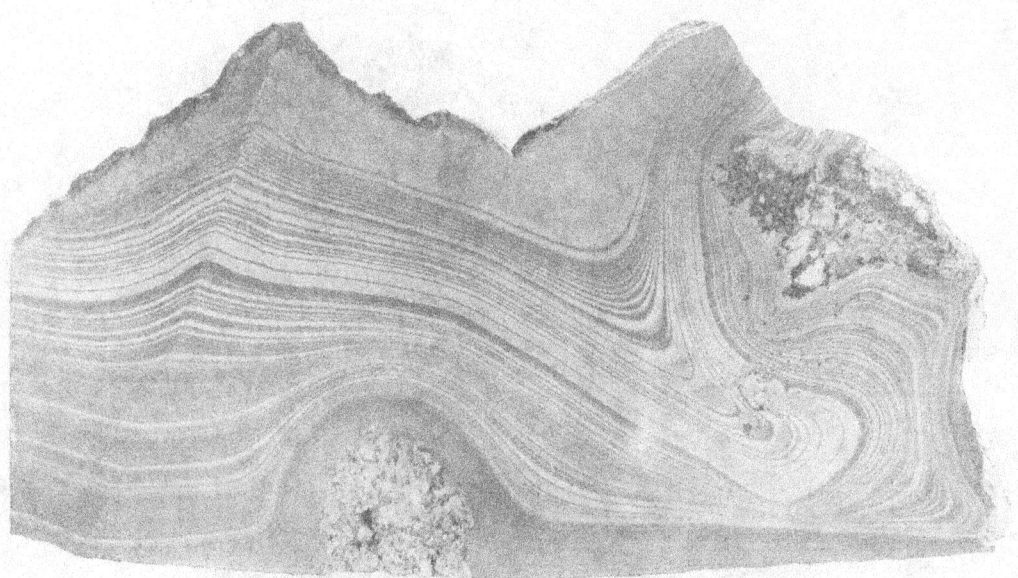

B. OXIDATION BANDING IN FINE, EVEN-GRAINED FELSITE.

Near North Star shaft. Natural size.

A.

B.

PHOTOMICROGRAPHS OF CONTACT BETWEEN OXIDIZED AND UNOXIDIZED PORTIONS OF
SPECIMEN SHOWN IN PLATE XIV, *A.*

A was taken in ordinary light and *B* in polarized light. In the unoxidized half, as shown at the left in *A*, small
pyrite crystals abound; in the oxidized half the iron occurs as finely divided oxide, though the spaces originally
occupied by pyrite are still recognizable. *B* shows that the oxidation bending is in a general way independent
of the original rock texture. It will be noted, however, that very little iron oxide has developed in the
phenocrysts.

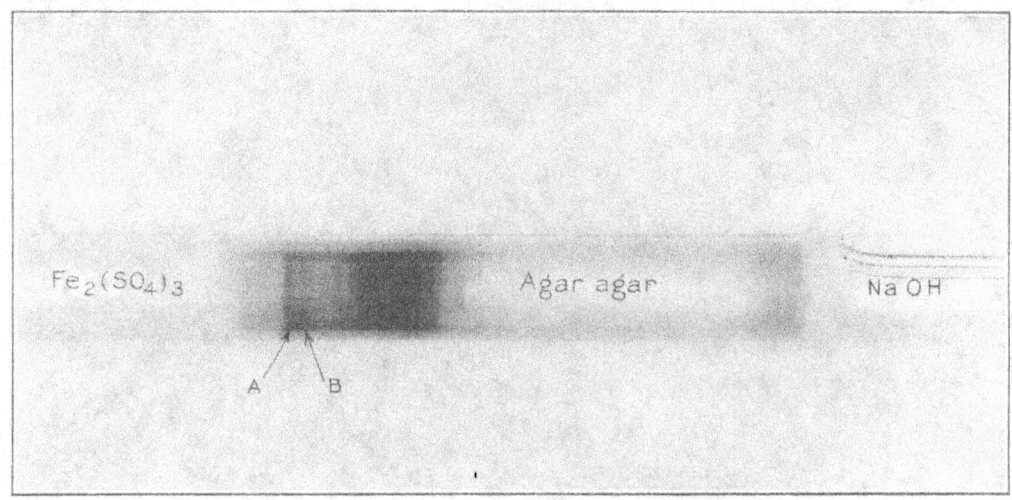

A.

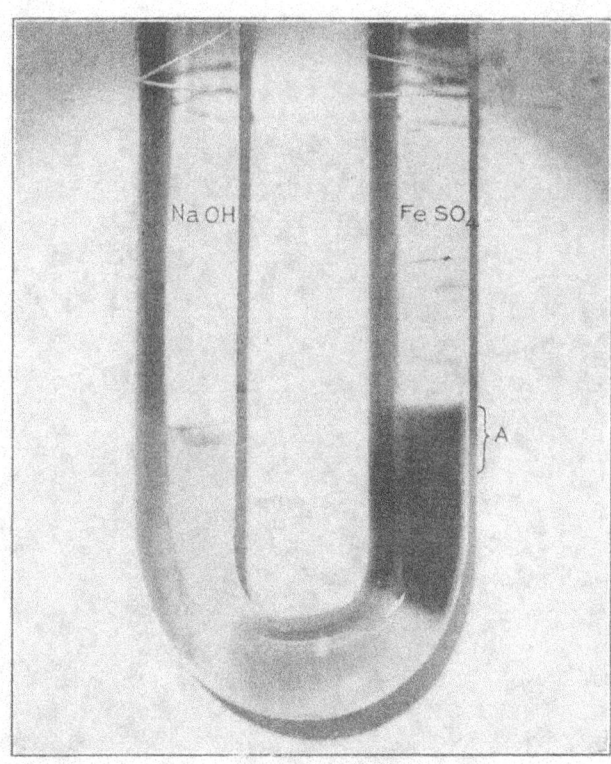

B.

LIESEGANG RINGS OF FERROUS AND FERRIC HYDROXIDES DEVELOPED IN AGAR AGAR BY
DIFFUSION FROM OPPOSITE ENDS OF TUBES OF IRON SULPHATES AND SODIUM HYDROXIDE.

A. After 17 days. A is the first and B the second ring formed.
B. After one month. Most of the rings were ferrous hydroxide, but in the late stages, when the iron solution
     had oxidized, rings of ferric hydroxide were developed at A.

GOLD RUSH BOOKS

OREGON, USA

www.GoldMiningBooks.com

Books On Mining

Visit: www.goldminingbooks.com to order your copies or ask your favorite book seller to offer them.

Mining Books by Kerby Jackson

<u>Gold Dust: Stories From Oregon's Mining Years</u> - Oregon mining historian and prospector, Kerby Jackson, brings you a treasure trove of seventeen stories on Southern Oregon's rich history of gold prospecting, the prospectors and their discoveries, and the breathtaking areas they settled in and made homes. 5" X 8", 98 ppgs. Retail Price: $11.99

<u>The Golden Trail: More Stories From Oregon's Mining Years</u> - In his follow-up to "Gold Dust: Stories of Oregon's Mining Years", this time around, Jackson brings us twelve tales from Oregon's Gold Rush, including the story about the first gold strike on Canyon Creek in Grant County, about the old timers who found gold by the pail full at the Victor Mine near Galice, how Iradel Bray discovered a rich ledge of gold on the Coquille River during the height of the Rogue River War, a tale of two elderly miners on the hunt for a lost mine in the Cascade Mountains, details about the discovery of the famous Armstrong Nugget and others. 5" X 8", 70 ppgs. Retail Price: $10.99

Oregon Mining Books

<u>Geology and Mineral Resources of Josephine County, Oregon</u> - Unavailable since the 1970's, this important publication was originally compiled by the Oregon Department of Geology and Mineral Industries and includes important details on the economic geology and mineral resources of this important mining area in South Western Oregon. Included are notes on the history, geology and development of important mines, as well as insights into the mining of gold, copper, nickel, limestone, chromium and other minerals found in large quantities in Josephine County, Oregon. 8.5" X 11", 54 ppgs. Retail Price: $9.99

<u>Mines and Prospects of the Mount Reuben Mining District</u> - Unavailable since 1947, this important publication was originally compiled by geologist Elton Youngberg of the Oregon Department of Geology and Mineral Industries and includes detailed descriptions, histories and the geology of the Mount Reuben Mining District in Josephine County, Oregon. Included are notes on the history, geology, development and assay statistics, as well as underground maps of all the major mines and prospects in the vicinity of this much neglected mining district. 8.5" X 11", 48 ppgs. Retail Price: $9.99

<u>The Granite Mining District</u> - Notes on the history, geology and development of important mines in the well known Granite Mining District which is located in Grant County, Oregon. Some of the mines discussed include the Ajax, Blue Ribbon, Buffalo, Continental, Cougar-Independence, Magnolia, New York, Standard and the Tillicum. Also included are many rare maps pertaining to the mines in the area. 8.5" X 11", 48 ppgs. Retail Price: $9.99

<u>Ore Deposits of the Takilma and Waldo Mining Districts of Josephine County, Oregon</u> - The Waldo and Takilma mining districts are most notable for the fact that the earliest large scale mining of placer gold and copper in Oregon took place in these two areas. Included are details about some of the earliest large gold mines in the state such as the Llano de Oro, High Gravel, Cameron, Platerica, Deep Gravel and others, as well as copper mines such as the famous Queen of Bronze mine, the Waldo, Lily and Cowboy mines. This volume also includes six maps and 20 original illustrations. 8.5" X 11", 74 ppgs. Retail Price: $9.99

<u>Metal Mines of Douglas, Coos and Curry Counties, Oregon</u> - Oregon mining historian Kerby Jackson introduces us to a classic work on Oregon's mining history in this important re-issue of Bulletin 14C Volume 1, otherwise known as the Douglas, Coos & Curry Counties, Oregon Metal Mines Handbook. Unavailable since 1940, this important publication was originally compiled by the Oregon Department of Geology and Mineral Industries includes detailed descriptions, histories and the geology of over 250 metallic mineral mines and prospects in this rugged area of South West Oregon. 8.5" X 11", 158 ppgs. Retail Price: $19.99

Metal Mines of Jackson County, Oregon - Unavailable since 1943, this important publication was originally compiled by the Oregon Department of Geology and Mineral Industries includes detailed descriptions, histories and the geology of over 450 metallic mineral mines and prospects in Jackson County, Oregon. Included are such famous gold mining areas as Gold Hill, Jacksonville, Sterling and the Upper Applegate. **8.5" X 11", 220 ppgs. Retail Price: $24.99**

Metal Mines of Josephine County, Oregon - Oregon mining historian Kerby Jackson introduces us to a classic work on Oregon's mining history in this important re-issue of Bulletin 14C, otherwise known as the Josephine County, Oregon Metal Mines Handbook. Unavailable since 1952, this important publication was originally compiled by the Oregon Department of Geology and Mineral Industries includes detailed descriptions, histories and the geology of over 500 metallic mineral mines and prospects in Josephine County, Oregon. **8.5" X 11", 250 ppgs. Retail Price: $24.99**

Metal Mines of North East Oregon - Oregon mining historian Kerby Jackson introduces us to a classic work on Oregon's mining history in this important re-issue of Bulletin 14A and 14B, otherwise known as the North East Oregon Metal Mines Handbook. Unavailable since 1941, this important publication was originally compiled by the Oregon Department of Geology and Mineral Industries and includes detailed descriptions, histories and the geology of over 750 metallic mineral mines and prospects in North Eastern Oregon. **8.5" X 11", 310 ppgs. Retail Price: $29.99**

Metal Mines of North West Oregon - Oregon mining historian Kerby Jackson introduces us to a classic work on Oregon's mining history in this important re-issue of Bulletin 14D, otherwise known as the North West Oregon Metal Mines Handbook. Unavailable since 1951, this important publication was originally compiled by the Oregon Department of Geology and Mineral Industries and includes detailed descriptions, histories and the geology of over 250 metallic mineral mines and prospects in North Western Oregon. **8.5" X 11", 182 ppgs. Retail Price: $19.99**

Mines and Prospects of Oregon - Mining historian Kerby Jackson introduces us to a classic mining work by the Oregon Bureau of Mines in this important re-issue of The Handbook of Mines and Prospects of Oregon. Unavailable since 1916, this publication includes important insights into hundreds of gold, silver, copper, coal, limestone and other mines that operated in the State of Oregon around the turn of the 19th Century. Included are not only geological details on early mines throughout Oregon, but also insights into their history, production, locations and in some cases, also included are rare maps of their underground workings. **8.5" X 11", 314 ppgs. Retail Price: $24.99**

Lode Gold of the Klamath Mountains of Northern California and South West Oregon
(See California Mining Books)

Mineral Resources of South West Oregon - Unavailable since 1914, this publication includes important insights into dozens of mines that once operated in South West Oregon, including the famous gold fields of Josephine and Jackson Counties, as well as the Coal Mines of Coos County. Included are not only geological details on early mines throughout South West Oregon, but also insights into their history, production and locations. **8.5" X 11", 154 ppgs. Retail Price: $11.99**

Chromite Mining in The Klamath Mountains of California and Oregon
(See California Mining Books)

Southern Oregon Mineral Wealth - Unavailable since 1904, this rare publication provides a unique snapshot into the mines that were operating in the area at the time. Included are not only geological details on early mines throughout South West Oregon, but also insights into their history, production and locations. Some of the mining areas include Grave Creek, Greenback, Wolf Creek, Jump Off Joe Creek, Granite Hill, Galice, Mount Reuben, Gold Hill, Galls Creek, Kane Creek, Sardine Creek, Birdseye Creek, Evans Creek, Foots Creek, Jacksonville, Ashland, the Applegate River, Waldo, Kerby and the Illinois River, Althouse and Sucker Creek, as well as insights into local copper mining and other topics. **8.5" X 11", 64 ppgs. Retail Price: $8.99**

Geology and Ore Deposits of the Takilma and Waldo Mining Districts - Unavailable since the 1933, this publication was originally compiled by the United States Geological Survey and includes details on gold and copper mining in the Takilma and Waldo Districts of Josephine County, Oregon. The Waldo and Takilma mining districts are most notable for the fact that the earliest large scale mining of placer gold and copper in Oregon took place in these two areas. Included in this report are details about some of the earliest large gold mines in the state such as the Llano de Oro, High Gravel, Cameron, Platerica, Deep Gravel and others, as well as copper mines such as the famous Queen of Bronze mine, the Waldo, Lily and Cowboy mines. In addition to geological examinations, insights are also provided into the production, day to day operations and early histories of these mines, as well as calculations of known mineral reserves in the area. This volume also includes six maps and 20 original illustrations. **8.5" X 11", 74 ppgs. Retail Price: $9.99**

Gold Mines of Oregon - Oregon mining historian Kerby Jackson introduces us to a classic work on Oregon's mining history in this important re-issue of Bulletin 61, otherwise known as "Gold and Silver In Oregon". Unavailable since 1968, this important publication was originally compiled by geologists Howard C. Brooks and Len Ramp of the Oregon Department of Geology and Mineral Industries and includes detailed descriptions, histories and the geology of over 450 gold mines Oregon. Included are notes on the history, geology and gold production statistics of all the major mining areas in Oregon including the Klamath Mountains, the Blue Mountains and the North Cascades. While gold is where you find it, as every miner knows, the path to success is to prospect for gold where it was previously found. 8.5" X 11", 344 ppgs. **Retail Price: $24.99**

Mines and Mineral Resources of Curry County Oregon - Originally published in 1916, this important publication on Oregon Mining has not been available for nearly a century. Included are rare insights into the history, production and locations of dozens of gold mines in Curry County, Oregon, as well as detailed information on important Oregon mining districts in that area such as those at Agness, Bald Face Creek, Mule Creek, Boulder Creek, China Diggings, Collier Creek, Elk River, Gold Beach, Rock Creek, Sixes River and elsewhere. Particular attention is especially paid to the famous beach gold deposits of this portion of the Oregon Coast. 8.5" X 11", 140 ppgs. **Retail Price: $11.99**

Chromite Mining in South West Oregon - Originally published in 1961, this important publication on Oregon Mining has not been available for nearly a century. Included are rare insights into the history, production and locations of nearly 300 chromite mines in South Western Oregon. 8.5" X 11", 184 ppgs. **Retail Price: $14.99**

Mineral Resources of Douglas County Oregon - Originally published in 1972, this important publication on Oregon Mining has not been available for nearly forty years. Included are rare insights into the geology, history, production and locations of numerous gold mines and other mining properties in Douglas County, Oregon. 8.5" X 11", 124 ppgs. **Retail Price: $11.99**

Mineral Resources of Coos County Oregon - Originally published in 1972, this important publication on Oregon Mining has not been available for nearly forty years. Included are rare insights into the geology, history, production and locations of numerous gold mines and other mining properties in Coos County, Oregon. 8.5" X 11", 100 ppgs. **Retail Price: $11.99**

Mineral Resources of Lane County Oregon - Originally published in 1938, this important publication on Oregon Mining has not been available for nearly seventy five years. Included are extremely rare insights into the geology and mines of Lane County, Oregon, in particular in the Bohemia, Blue River, Oakridge, Black Butte and Winberry Mining Districts. 8.5" X 11", 82 ppgs. **Retail Price: $9.99**

Mineral Resources of the Upper Chetco River of Oregon: Including the Kalmiopsis Wilderness - Originally published in 1975, this important publication on Oregon Mining has not been available for nearly forty years. Withdrawn under the 1872 Mining Act since 1984, real insight into the minerals resources and mines of the Upper Chetco River has long been unavailable due to the remoteness of the area. Despite this, the decades of battle between property owners and environmental extremists over the last private mining inholding in the area has continued to pique the interest of those interested in mining and other forms of natural resource use. Gold mining began in the area in the 1850's and has a rich history in this geographic area, even if the facts surrounding it are little known. Included are twenty two rare photographs, as well as insights into the Becca and Morning Mine, the Emmly Mine (also known as Emily Camp), the Frazier Mine, the Golden Dream or Higgins Mine, Hustis Mine, Peck Mine and others. 8.5" X 11", 64 ppgs. **Retail Price: $8.99**

Gold Dredging in Oregon - Originally published in 1939, this important publication on Oregon Mining has not been available for nearly seventy five years. Included are extremely rare insights into the history and day to day operations of the dragline and bucketline gold dredges that once worked the placer gold fields of South West and North East Oregon in decades gone by. Also included are details into the areas that were worked by gold dredges in Josephine, Jackson, Baker and Grant counties, as well as the economic factors that impacted this mining method. This volume also offers a unique look into the values of river bottom land in relation to both farming and mining, in how farm lands were mined, re-soiled and reclamated after the dredges worked them. Featured are hard to find maps of the gold dredge fields, as well as rare photographs from a bygone era. 8.5" X 11", 86 ppgs. **Retail Price: $8.99**

Quick Silver Mining in Oregon - Originally published in 1963, this important publication on Oregon Mining has not been available for over fifty years. This publication includes details into the history and production of Elemental Mercury or Quicksilver in the State of Oregon. 8.5" X 11", 238 ppgs. **Retail Price: $15.99**

Mines of the Greenhorn Mining District of Grant County Oregon - Originally published in 1948, this important publication on Oregon Mining has not been available for over sixty five years. In this publication are rare insights into the mines of the famous Greenhorn Mining District of Grant County, Oregon, especially the famous Morning Mine. Also included are details on the Tempest, Tiger, Bi-Metallic, Windsor, Psyche, Big Johnny, Snow Creek, Banzette and Paramount Mines, as well as prospects in the vicinities in the famous mining areas of Mormon Basin, Vinegar Basin and Desolation Creek. Included are hard to find mine maps and dozens of rare photographs from the bygone era of Grant County's rich mining history. 8.5" X 11", 72 ppgs. **Retail Price: $9.99**

Geology of the Wallowa Mountains of Oregon: Part I (Volume 1) - Originally published in 1938, this important publication on Oregon Mining has not been available for nearly seventy five years. Included are details on the geology of this unique portion of North Eastern Oregon. This is the first part of a two book series on the area. Accompanying the text are rare photographs and historic maps.**8.5″ X 11″, 92 ppgs. Retail Price: $9.99**

Geology of the Wallowa Mountains of Oregon: Part II (Volume 2) - Originally published in 1938, this important publication on Oregon Mining has not been available for nearly seventy five years. Included are details on the geology of this unique portion of North Eastern Oregon. This is the first part of a two book series on the area. Accompanying the text are rare photographs and historic maps.**8.5″ X 11″, 94 ppgs. Retail Price: $9.99**

Field Identification of Minerals For Oregon Prospectors - Originally published in 1940, this important publication on Oregon Mining has not been available for nearly seventy five years. Included in this volume is an easy system for testing and identifying a wide range of minerals that might be found by prospectors, geologists and rockhounds in the State of Oregon, as well as in other locales. Topics include how to put together your own field testing kit and how to conduct rudimentary tests in the field. This volume is written in a clear and concise way to make it useful even for beginners. **8.5″ X 11″, 158 ppgs. Retail Price: $14.99**

The Bohemia Mining District of Oregon - Originally published in 1900, this important publication on Oregon Mining has not been available for over a century. Included in this volume are important insights into the famous Bohemia Mining District of Oregon, including the histories and locations of important gold mines in the area such as the Ophir Mine, Clarence, Acturas, Peek-a-boo, White Swan, Combination Mine, the Musick Mine, The California, White Ghost, The Mystery, Wall Street, Vesuvius, Story, Lizzie Bullock, Delta, Elsie Dora, Golden Slipper, Broadway, Champion Mine, Knott, Noonday, Helena, White Wings, Riverside and others. Also included are notes on the nearby Blue River Mining District. **8.5″ X 11″, 58 ppgs. Retail Price: $9.99**

The Gold Fields of Eastern Oregon - Unavailable since 1900, this publication was originally compiled by the Baker City Chamber of Commerce Offering important insights into the gold mining history of Eastern Oregon, "The Gold Fields of Eastern Oregon" sheds a rare light on many of the gold mines that were operating at the turn of the 19th Century in Baker County and Grant County in North Eastern Oregon. Some of the areas featured include the Cable Cove District, Baisely-Elhorn, Granite, Red Boy, Bonanza, Susanville, Sparta, Virtue, Vaughn, Sumpter, Burnt River, Rye Valley and other mining districts. Included is basic information on not only many gold mines that are well known to those interested in Eastern Oregon mining history, but also many mines and prospects which have been mostly lost to the passage of time. Accompanying are numerous rare photos **8.5″ X 11″, 78 ppgs. Retail Price: $10.99**

Gold Mining in Eastern Oregon - Originally published in 1938, this important publication on Oregon Mining has not been available for over a century. Included in this volume are important insights into the famous mining districts of Eastern Oregon during the late 1930's. Particular attention is given to those gold mines with milling and concentrating facilities in the Greenhorn, Red Boy, Alamo, Bonanza, Granite, Cable Cove, Cracker Creek, Virtue, Keating, Medical Springs, Sanger, Sparta, Chicken Creek, Mormon Basin, Connor Creek, Cornucopia and the Bull Run Mining Districts. Some of the mines featured include the Ben Harrison, North Pole-Columbia, Highland Maxwell, Baisley-Elkhorn, White Swan, Balm Creek, Twin Baby, Gem of Sparta, New Deal, Gleason, Gifford-Johnson, Cornucopia, Record, Bull Run, Orion and others. Of particular interest are the mill flow sheets and descriptions of milling operations of these mines. **8.5″ X 11″, 68 ppgs. Retail Price: $8.99**

The Gold Belt of the Blue Mountains of Oregon - Originally published in 1901, this important publication on Oregon Mining has not been available for over a century. Included in this volume are rare insights into the gold deposits of the Blue Mountains of North East Oregon, including the history of their early discovery and early production. Extensive details are offered on this important mining area's mineralogy and economic geology, as well as insights into nearby gold placers, silver deposits and copper deposits. Featured are the Elkhorn and Rock Creek mining districts, the Pocahontas district, Auburn and Minersville districts, Sumpter and Cracker Creek, Cable Cove, the Camp Carson district, Granite, Alamo, Greenhorn, Robinsonville, the Upper Burnt River Valley and Bonanza districts, Susanville, Quartzburg, Canyon Creek, Virtue, the Copper Butte district, the North Powder River, Sparta, Eagle Creek, Cornucopia, Pine Creek, Lower Powder River, the Upper Snake River Canyon, Rye Valley, Lower Burnt River Valley, Mormon Basin, the Malheur and Clarks Creek districts, Sutton Creek and others. Of particular interest are important details on numerous gold mines and prospects in these mining districts, including their locations, histories, geology and other important information, as well as information on silver, copper and fire opal deposits. **8.5″ X 11″, 250 ppgs. Retail Price: $24.99**

Mining in the Cascades Range of Oregon - Originally published in 1938, this important publication on Oregon Mining has not been available for over seventy five years. Included in this volume are rare insights into the gold mines and other types of metal mines in the Cascades Mountain Range of Oregon. Some of the important mining areas covered include the famous Bohemia Mining District, the North Santiam Mining District, Quartzville Mining District, Blue River Mining District, Fall Creek Mining District, Oakridge District, Zinc District, Buzzard-Al Sarena District, Grand Cove, Climax District and Barron Mining District. Of particular interest are important details on over 100 mines and prospects in these mining districts, including their locations, histories, geology and other important information. **8.5" X 11", 170 ppgs. Retail Price: $14.99**

Beach Gold Placers of the Oregon Coast - Originally published in 1934, this important publication on Oregon Mining has not been available for over 80 years. Included in this volume are rare insights into the beach gold deposits of the State of Oregon, including their locations, occurance, composition and geology. Of particular interest is information on placer platinum in Oregon's rich beach deposits. Also included are the locations and other information on some famous Oregon beach mines, including the Pioneer, Eagle, Chickamin, Iowa and beach placer mines north of the mouth of the Rogue River. **8.5" X 11", 60 ppgs. Retail Price: $8.99**

Idaho Mining Books

Gold in Idaho - Unavailable since the 1940's, this publication was originally compiled by the Idaho Bureau of Mines and includes details on gold mining in Idaho. Included is not only raw data on gold production in Idaho, but also valuable insight into where gold may be found in Idaho, as well as practical information on the gold bearing rocks and other geological features that will assist those looking for placer and lode gold in the State of Idaho. This volume also includes thirteen gold maps that greatly enhance the practical usability of the information contained in this small book detailing where to find gold in Idaho. **8.5" X 11", 72 ppgs. Retail Price: $9.99**

Geology of the Couer D'Alene Mining District of Idaho - Unavailable since 1961, this publication was originally compiled by the Idaho Bureau of Mines and Geology and includes details on the mining of gold, silver and other minerals in the famous Coeur D'Alene Mining District in Northern Idaho. Included are details on the early history of the Coeur D'Alene Mining District, local tectonic settings, ore deposit features, information on the mineral belts of the Osburn Fault, as well as detailed information on the famous Bunker Hill Mine, the Dayrock Mine, Galena Mine, Lucky Friday Mine and the infamous Sunshine Mine. This volume also includes sixteen hard to find maps. **8.5" X 11", 70 ppgs. Retail Price: $9.99**

The Gold Camps and Silver Cities of Idaho - Originally published in 1963, this important publication on Idaho Mining has not been available for nearly fifty years. Included are rare insights into the history of Idaho's Gold Rush, as well as the mad craze for silver in the Idaho Panhandle. Documented in fine detail are the early mining excitements at Boise Basin, at South Boise, in the Owyhees, at Deadwood, Long Valley, Stanley Basin and Robinson Bar, at Atlanta, on the famous Boise River, Volcano, Little Smokey, Banner, Boise Ridge, Hailey, Leesburg, Lemhi, Pearl, at South Mountain, Shoup and Ulysses, Yellow Jacket and Loon Creek. The story follows with the appearance of Chinese miners at the new mining camps on the Snake River, Black Pine, Yankee Fork, Bay Horse, Clayton, Heath, Seven Devils, Gibbonsville, Vienna and Sawtooth City. Also included are special sections on the Idaho Lead and Silver mines of the late 1800's, as well as the mining discoveries of the early 1900's that paved the way for Idaho's modern mining and mineral industry. Lavishly illustrated with rare historic photos, this volume provides a one of a kind documentary into Idaho's mining history that is sure to be enjoyed by not only modern miners and prospectors who still scour the hills in search of nature's treasures, but also those enjoy history and tromping through overgrown ghost towns and long abandoned mining camps. **8.5" X 11", 186 ppgs. Retail Price: $14.99**

Ore Deposits and Mining in North Western Custer County Idaho - Unavailable since 1913, this important publication was originally published by the Us Department of the Interior and has been unavailable for a century. Included are fine details on the geology, geography, gold placers and gold and silver bearing quartz veins of the mining region of North West Custer County, Idaho. Of particular interest is a rare look at the mines and prospects of the region, including those such as the Ramshorn Mine, SkyLark, Riverview, Excelsior, Beardsley, Pacific, Hoosier, Silver Brick, Forest Rose and dozens of others in the Bay Horse Mining District. Also covered are the mines of the Yankee Fork District such as the Lucky Boy, Badger, Black, Enterprise, Charles Dickens, Morrison, Golden Sunbeam, Montana, Golden Gate and others, as well as those in the Loon Mining District. **8.5" X 11", 126 ppgs. Retail Price: $12.99**

<u>**Gold Rush To Idaho**</u> - Unavailable since 1963, this important publication was originally published by the Idaho Bureau of Mines and has been unavailable for 50 years. "Gold Rush To Idaho" revisits the earliest years of the discovery of gold in Idaho Territory and introduces us to the conditions that the pioneer gold seekers met when they blazed a trail through the wilderness of Idaho's mountains and discovered the precious yellow metal at Oro Fino and Pierce. Subsequent rushes followed at places like Elk City, Newsome, Clearwater Station, Florence, Warrens and elsewhere. Of particular interest is a rare look at the hardships that the first miners in Idaho met with during their day to day existences and their attempts to bring law and order to their mining camps. **8.5" X 11", 88 ppgs. Retail Price: $9.99**

<u>**The Geology and Mines of Northern Idaho and North Western Montana**</u> - Unavailable since 1909, this important publication was originally published by the Us Department of the Interior and has been unavailable for a century. Included are fine details on the geology and geography of the mining regions of Northern Idaho and North Western Montana. Of particular interest is a rare look at the mines and prospects of the region, including those in the Pine Creek Mining District, Lake Pend Oreille district, Troy Mining District, Sylvanite District, Cabinet Mining District, Prospect Mining District and the Missoula Valley. Some of the mines featured include the Iron Mountain, Silver Butte, Snowshoe, Grouse Mountain Mine and others. **8.5" X 11", 142 ppgs. Retail Price: $12.99**

<u>**Mining in the Alturas Quadrangle of Blaine County Idaho**</u> - Unavailable since 1922, this important publication was originally published by the Idaho Bureau of Mines and has been unavailable for ninety years. Topics include the geology, rock formations and the formation of ore deposits in this important mining area of Idaho. Of particular focus is information on the local geology, quartz veins and ore deposits of this portion of Idaho. Included are hard to find details, including the descriptions and locations of numerous gold and silver mines in the area including the Silver King, Pilgrim, Columbia, Lone Jack, Sunbeam, Pride of the West, Lucky Boy, Scotia, Atlanta, Beaver-Bidwell and others mines and prospects. **8.5" X 11", 56 ppgs. Retail Price: $8.99**

<u>**Mining in Lemhi County Idaho**</u> - Originally published in 1913, this important book on Idaho Mining has not been available to miners for over a century. Included are rare insights into hundreds of gold, silver, copper and other mines in this famous Idaho mining area. Details include the locations, geology, history, production and other facts of the mines of this region, not only gold and silver hardrock mines, but also gold placer mines, lead-silver deposits, copper mines, cobalt-nickel deposits, tungsten and tin mines . It is lavishly illustrated with hard to find photos of the period and rare mining maps. Some of the vicinities featured include the Nicholia Mining District, Spring Mountain District, Texas District, Blue Wing District, Junction District, McDevitt District, Pratt Creek, Eldorado District, Kirtley Creek, Carmen Creek, Gibbonsville, Indian Creek, Mineral Hill District, Mackinaw, Eureka District, Blackbird District, YellowJacket District, Gravel Range District, Junction District, Parker Mountain and other mining districts. **8.5" X 11", 226 ppgs. Retail Price: $19.99**

Utah Mining Books

<u>**Fluorite in Utah**</u> - Unavailable since 1954, this publication was originally compiled by the USGS, State of Utah and U.S. Atomic Energy Commission and details the mining of fluorspar, also known as fluorite in the State of Utah. Included are details on the geology and history of fluorspar (fluorite) mining in Utah, including details on where this unique gem mineral may be found in the State of Utah. **8.5" X 11", 60 ppgs. Retail Price: $8.99**

California Mining Books

<u>**The Tertiary Gravels of the Sierra Nevada of California**</u> - Mining historian Kerby Jackson introduces us to a classic mining work by Waldemar Lindgren in this important re-issue of The Tertiary Gravels of the Sierra Nevada of California. Unavailable since 1911, this publication includes details on the gold bearing ancient river channels of the famous Sierra Nevada region of California. **8.5" X 11", 282 ppgs. Retail Price: $19.99**

<u>**The Mother Lode Mining Region of California**</u> - Unavailable since 1900, this publication includes details on the gold mines of California's famous Mother Lode gold mining area. Included are details on the geology, history and important gold mines of the region, as well as insights into historic mining methods, mine timbering, mining machinery, mining bell signals and other details on how these mines operated. Also included are insights into the gold mines of the California Mother Lode that were in operation during the first sixty years of California's mining history. **8.5" X 11", 176 ppgs. Retail Price: $14.99**

<u>**Lode Gold of the Klamath Mountains of Northern California and South West Oregon**</u> - Unavailable since 1971, this publication was originally compiled by Preston E. Hotz and includes details on the lode mining districts of Oregon and California's Klamath Mountains. Included are details on the geology, history and important lode mines of the French Gulch, Deadwood, Whiskeytown, Shasta, Redding, Muletown, South Fork, Old Diggings, Dog Creek (Delta), Bully Choop (Indian Creek), Harrison Gulch, Hayfork, Minersville, Trinity Center, Canyon Creek, East Fork, New River, Denny, Liberty (Black Bear), Cecilville, Callahan, Yreka, Fort Jones and Happy Camp mining districts in California, as well as the Ashland, Rogue River, Applegate, Illinois River, Takilma, Greenback, Galice, Silver Peak, Myrtle Creek and Mule Creek districts of South Western Oregon. Also included are insights into the mineralization and other characteristics of this important mining region. **8.5" X 11", 100 ppgs. Retail Price: $10.99**

Mines and Mineral Resources of Shasta County, Siskiyou County, Trinity County: California - Unavailable since 1915, this publication was originally compiled by the California State Mining Bureau and includes details on the gold mines of this area of Northern California. Also included are insights into the mineralization and other characteristics of this important mining region, as well as the location of historic gold mines. **8.5″ X 11″, 204 ppgs. Retail Price: $19.99**

Geology of the Yreka Quadrangle, Siskiyou County, California - Unavailable since 1977, this publication was originally compiled by Preston E. Hotz and includes details on the geology of the Yreka Quadrangle of Siskiyou County, California. Also included are insights into the mineralization and other characteristics of this important mining region. **8.5″ X 11″, 78 ppgs. Retail Price: $7.99**

Mines of San Diego and Imperial Counties, California - Originally published in 1914, this important publication on California Mining has not been available for a century. This publication includes important information on the early gold mines of San Diego and Imperial County, which were some of the first gold fields mined in California by early Spanish and Mexican miners before the 49ers came on the scene. Included are not only details on early mining methods in the area, production statistics and geological information, but also the location of the early gold mines that helped make California "The Golden State". Also included are details on the mining of other minerals such as silver, lead, zinc, manganese, tungsten, vanadium, asbestos, barite, borax, cement, clay, dolomite, fluospar, gem stones, graphite, marble, salines, petroleum, stronium, talc and others. **8.5″ X 11″, 116 ppgs. Retail Price: $12.99**

Mines of Sierra County, California - Unavailable since 1920, this publication was originally compiled by the California State Mining Bureau and includes details on the gold mines of Sierra County, California. Also included are insights into the mineralization and other characteristics of this important mining region, as well as the location of historic gold mines. **8.5″ X 11″, 156 ppgs. Retail Price: $19.99**

Mines of Plumas County, California - Unavailable since 1918, this publication was originally compiled by the California State Mining Bureau and includes details on the gold mines of Plumas County, California. Also included are insights into the mineralization and other characteristics of this important mining region, as well as the location of historic gold mines. **8.5″ X 11″, 200 ppgs. Retail Price: $19.99**

Mines of El Dorado, Placer, Sacramento and Yuba Counties, California - Originally published in 1917, this important publication on California Mining has not been available for nearly a century. This publication includes important information on the early gold mines of El Dorado County, Placer County, Sacramento County and Yuba County, which were some of the first gold fields mined by the Forty-Niners during the California Gold Rush. Included are not only details on early mining methods in the area, production statistics and geological information, but also the location of the early gold mines that helped make California "The Golden State". Also included are insights into the early mining of chrome, copper and other minerals in this important mining area. **8.5″ X 11″, 204 ppgs. Retail Price: $19.99**

Mines of Los Angeles, Orange and Riverside Counties, California - Originally published in 1917, this important publication on California Mining has not been available for nearly a century. This publication includes important information on the early gold mines of Los Angeles County, Orange County and Riverside County, which were some of the first gold fields mined in California by early Spanish and Mexican miners before the 49ers came on the scene. Included are not only details on early mining methods in the area, production statistics and geological information, but also the location of the early gold mines that helped make California "The Golden State". **8.5″ X 11″, 146 ppgs. Retail Price: $12.99**

Mines of San Bernadino and Tulare Counties, California - Originally published in 1917, this important publication on California Mining has not been available for nearly a century. This publication includes important information on the early gold mines of San Bernadino and Tulare County, which were some of the first gold fields mined in California by early Spanish and Mexican miners before the 49ers came on the scene. Included are not only details on early mining methods in the area, production statistics and geological information, but also the location of the early gold mines that helped make California "The Golden State". Also included are details on the mining of other minerals such as copper, iron, lead, zinc, manganese, tungsten, vanadium, asbestos, barite, borax, cement, clay, dolomite, fluospar, gem stones, graphite, marble, salines, petroleum, stronium, talc and others. **8.5″ X 11″, 200 ppgs. Retail Price: $19.99**

Chromite Mining in The Klamath Mountains of California and Oregon - Unavailable since 1919, this publication was originally compiled by J.S. Diller of the United States Department of Geological Survey and includes details on the chromite mines of this area of Northern California and Southern Oregon. Also included are insights into the mineralization and other characteristics of this important mining region, as well as the location of historic mines. Also included are insights into chromite mining in Eastern Oregon and Montana. **8.5″ X 11″, 98 ppgs. Retail Price: $9.99**

Mines and Mining in Amador, Calaveras and Tuolumne Counties, California - Unavailable since 1915, this publication was originally compiled by William Tucker and includes details on the mines and mineral resources of this important California mining area. Included are details on the geology, history and important gold mines of the region, as well as insights into other local mineral resources such as asbestos, clay, copper, talc, limestone and others. Also included are insights into the mineralization and other characteristics of this important portion of California's Mother Lode mining region. **8.5" X 11", 198 ppgs. Retail Price: $14.99**

The Cerro Gordo Mining District of Inyo County California - Unavailable since 1963, this publication was originally compiled by the United States Department of Interior. Included are insights into the mineralization and other characteristics of this important mining region of Southern California. Topics include the mining of gold and silver in this important mining district in Inyo County, California, including details on the history, production and locations of the Cerro Gordo Mine, the Morning Star Mine, Estelle Tunnel, Charles Lease Tunnel, Ignacio, Hart, Crosscut Tunnel, Sunset, Upper Newtown, Newtown, Ella, Perseverance, Newsboy, Belmont and other silver and gold mines in the Cerro Gordo Mining District. This volume also includes important insights into the fossil record, geologic formations, faults and other aspects of economic geology in this California mining district. **8.5" X 11", 104 ppgs. Retail Price: $10.99**

Mining in Butte, Lassen, Modoc, Sutter and Tehama Counties of California - Unavailable since 1917, this publication was originally compiled by the United States Department of Interior. Included are insights into the mineralization and other characteristics of this important mining region of California. Topics include the mining of asbestos, chromite, gold, diamonds and manganese in Butte County, the mining of gold and copper in the Hayden Hill and Diamond Mountain mining districts of Lassen County, the mining of coal, salt, copper and gold in the High Grade and Winters mining districts of Modoc County, gold mining in Sutter County and the mining of gold, chromite, manganese and copper in Tehama County. This volume also includes the production records and locations of numerous mines in this important mining region. **8.5" X 11", 114 ppgs. Retail Price: $11.99**

Mines of Trinity County California - Originally published in 1965, this important publication on California Mining has not been available for nearly fifty years. This publication includes important information on mines and mining in Trinity County, California, as well insights into the mineralization and geology of this important mining area in Northern California. Included are extensive details on hardrock and placer gold mines and prospects, including charts showing the locations of these historic mines.. **8.5" X 11", 144 ppgs. Retail Price: $12.99**

Mines of Kern County California - Originally published in 1962, this important publication on California Mining has not been available for nearly fifty years. This publication includes important information on mines and mining in Kern County, California, as well insights into the mineralization and geology of this important mining area in California. Included are extensive details on hardrock and placer gold mines and prospects, including charts showing the locations of these historic mines. **8.5" X 11", 398 ppgs. Retail Price: $24.99**

Mines of Calaveras County California - Originally published in 1962, this important publication on California Mining has not been available for nearly fifty years. This publication includes important information on mines and mining in Calaveras County, California, as well insights into the mineralization and geology of this important mining area in Northern California. Included are extensive details on hardrock and placer gold mines and prospects, including charts showing the locations of these historic mines. **8.5" X 11", 236 ppgs. Retail Price: $19.99**

Lode Gold Mining in Grass Valley California - Unavailable since 1940, this publication was originally compiled by the United States Department of Interior. Included are insights into the gold mineralization and other characteristics of this important mining region of Nevada County, California. This volume also includes important insights into the geologic formations, faults and other aspects of economic geology in this California mining district. Of particular interest are the fine details on many hardrock gold mines in the area, including their locations, histories, development and mineralization. Some of the mines featured include the Gold Hill Mine, Massachusetts Hill, Boundary, Peabody, Golden Center, North Star, Omaha, Lone Jack, Homeward Bound, Hartery, Wisconsin, Allison Ranch, Phoenix, Kate Hayes, W.Y.O.D., Empire, Rich Hill, Daisy Hill, Orleans, Sultana, Centennial, Conlin, Ben Franklin, Crown Point and many others. **8.5" X 11", 148 ppgs. Retail Price: $12.99**

Lode Mining in the Alleghany District of Sierra County California - Unavailable since 1913, this publication was originally compiled by the United States Department of Interior. Included are insights into the mineralization and other characteristics of this important mining region of Sierra County. Included are details on the history, production and locations of numerous hardrock gold mines in this famous California area, including the Tightner Mine, Minnie D., Osceola, Eldorado, Twenty One, Sherman, Kenton, Oriental, Rainbow, Plumbago, Irelan, Gold Canyon, North Fork, Federal, Kate Hardy and others. This volume also includes important insights into the fossil record, geologic formations, faults and other aspects of economic geology in this California mining district. **8.5" X 11", 48 ppgs. Retail Price: $7.99**

Six Months In The Gold Mines During The California Gold Rush - Unavailable since 1850, this important work is a first hand account of one "49'ers" personal experience during the great California Gold Rush, shedding important light on one of the most exciting periods in the history of not only California, but also the world. Compiled from journals written between 1847 and 1849 by E. Gould Buffum, a native of New York, "Six Months In The Gold Mines During The California Gold Rush" offers a rare look into the day to day lives of the people who came to California to work in her gold mines when the state was still a great frontier. **8.5" X 11", 290 ppgs. Retail Price: $19.99**

Quartz Mines of the Grass Valley Mining District of California - Unavailable since 1867, this important publication has not been available since those days. This rare publication offers a short dissertation on the early hardrock mines in this important mining district in the California Mother Lode region between the 1850's and 1860's. Also included are hard to find details on the mineralization and locations of these mines, as well as how they were operated in those day. **8.5" X 11", 44 ppgs. Retail Price: $8.99**

Alaska Mining Books

Ore Deposits of the Willow Creek Mining District, Alaska - Unavailable since 1954, this hard to find publication includes valuable insights into the Willow Creek Mining District near Hatcher Pass in Alaska. The publication includes insights into the history, geology and locations of the well known mines in the area, including the Gold Cord, Independence, Fern, Mabel, Lonesome, Snowbird, Schroff-O'Neil, High Grade, Marion Twin, Thorpe, Webfoot, Kelly-Willow, Lane, Holland and others. **8.5" X 11", 96 ppgs. Retail Price: $9.99**

The Juneau Gold Belt of Alaska - Unavailable since 1906, this hard to find publication includes valuable insights into the gold mines around Juneau, Alaska. The publication includes important details into the history, geology and locations of the well known gold mines and prospects in the area, including those around Windham Bay, Holkham Bay, Port Snettisham, on Grindstone and Rhine Creeks, Gold Creek, Douglas Island, Salmon Creek, Lemon Creek, Nugget Creek, from the Mendenhall River to Berners Bay, McGinnis Creek, Montana Creek, Peterson Creek, Windfall Creek, the Eagle River, Yankee Basin, Yankee Curve, Kowee Creek and elsewhere. Not only are gold placer mines included, but also hardrock gold mines. **8.5" X 11", 224 ppgs. Retail Price: $19.99**

Arizona Mining Books

Mines and Mining in Northern Yuma County Arizona - Originally published in 1911, this important publication on Arizona Mining has not been available for over a hundred years. Included are rare insights into the gold, silver, copper and quicksilver mines of Yuma County, Arizona together with hard to find maps and photographs. Some of the mines and mining districts featured include the Planet Copper Mine, Mineral Hill, the Clara Consolidated Mine, Viati Mine, Copper Basin prospect, Bowman Mine, Quartz King, Billy Mack, Carnation, the Wardwell and Osbourne, Valensuella Copper, the Mariquita, Colonial Mine, the French American, the New York-Plomosa, Guadalupe, Lead Camp, Mudersbach Copper Camp, Yellow Bird, the Arizona Northern (Salome Strike), Bonanza (Harqua Hala), Golden Eagle, Hercules, Socorro and others. **8.5" X 11", 144 ppgs. Retail Price: $11.99**

The Aravaipa and Stanley Mining Districts of Graham County Arizona - Originally published in 1925, this important publication on Arizona Mining has not been available for nearly ninety years. Included are rare insights into the gold and silver mines of these two important mining districts, together with hard to find maps. **8.5" X 11", 140 ppgs. Retail Price: $11.99**

Gold in the Gold Basin and Lost Basin Mining Districts of Mohave County, Arizona - This volume contains rare insights into the geology and gold mineralization of the Gold Basin and Lost Basin Mining Districts of Mohave County, Arizona that will be of benefit to miners and prospectors. Also included is a significant body of information on the gold mines and prospects of this portion of Arizona. This volume is lavishly illustrated with rare photos and mining maps. **8.5" X 11", 188 ppgs. Retail Price: $19.99**

Mines of the Jerome and Bradshaw Mountains of Arizona - This important publication on Arizona Mining has not been available for ninety years. This volume contains rare insights into the geology and ore deposits of the Jerome and Bradshaw Mountains of Arizona that will be of benefit to miners and prospectors who work those areas. Included is a significant body of information on the mines and prospects of the Verde, Black Hills, Cherry Creek, Prescott, Walker, Groom Creek, Hassayampa, Bigbug, Turkey Creek, Agua Fria, Black Canyon, Peck, Tiger, Pine Grove, Bradshaw, Tintop, Humbug and Castle Creek Mining Districts. This volume is lavishly illustrated with rare photos and mining maps. **8.5" X 11", 218 ppgs. Retail Price: $19.99**

The Ajo Mining District of Pima County Arizona - This important publication on Arizona Mining has not been available for nearly seventy years. This volume contains rare insights into the geology and mineralization of the Ajo Mining District in Pima County, Arizona and in particular the famous New Cornelia Mine. **8.5" X 11", 126 ppgs. Retail Price: $11.99**

<u>Mining in the Santa Rita and Patagonia Mountains of Arizona</u> - Originally published in 1915, this important publication on Arizona Mining has not been available for nearly a century. Included are rare insights into hundreds of gold, silver, copper and other mines in this famous Arizona mining area. Details include the locations, geology, history, production and other facts of the mines of this region. **8.5″ X 11″, 394 ppgs. Retail Price: $24.99**

<u>Mining in the Bisbee Quadrangle of Arizona</u> - Originally published in 1906, this important publication on Arizona Mining has not been available for nearly a century. Included are rare insights into hundreds of gold, silver, copper and other mines in this famous Arizona mining area. Details include the locations, geology, history, production and other facts of the mines of this important mining region. **8.5″ X 11″, 188 ppgs. Retail Price: $14.99**

Montana Mining Books

<u>A History of Butte Montana: The World's Greatest Mining Camp</u> - First published in 1900 by H.C. Freeman, this important publication sheds a bright light on one of the most important mining areas in the history of The West. Together with his insights, as well as rare photographs of the periods, Harry Freeman describes Butte and its vicinity from its early beginnings, right up to its flush years when copper flowed from its mines like a river. At the time of publication, Butte, Montana was known worldwide as "The Richest Mining Spot On Earth" and produced not only vast amounts of copper, but also silver, gold and other metals from its mines. Freeman illustrates, with great detail, the most important mines in the vicinity of Butte, providing rare details on their owners, their history and most importantly, how the mines operated and how their treasures were extracted. Of particular interest are the dozens of rare photographs that depict mines such as the famous Anaconda, the Silver Bow, the Smoke House, Moose, Paulin, Buffalo, Little Minah, the Mountain Consolidated, West Greyrock, Cora, the Green Mountain, Diamond, Bell, Parnell, the Neversweat, Nipper, Original and many others. **8.5″ X 11″, 142 ppgs. Retail Price: $12.99**

<u>The Butte Mining District of Montana</u> - This important publication on Montana Mining has not been available for over a century. Included are rare insights into the gold, copper and silver mines of Butte, Montana together with hard to find maps and photographs. Some of the topics include the early history of gold, silver and copper mining in the Butte area, insight into the geology of its mining areas, the local distribution of gold, silver and copper ores, as well their composition and how to identify them. Also included are detailed facts about the mines in the Butte Mining District, including the famous Anaconda Mine, Gagnon, Parrot, Blue Vein, Moscow, Poulin, Stella, Buffalo, Green Mountain, Wake Up Jim, the Diamond-Bell Group, Mountain Consolidated, East Greyrock, West Greyrock, Snowball, Corra, Speculator, Adirondack, Miners Union, the Jessie-Edith May Group, Otisco, Iduna, Colorado, Lizzie, Cambers, Anderson, Hesperus, Preferencia and dozens of others. **8.5″ X 11″, 298 ppgs. Retail Price: $24.99**

<u>Mines of the Helena Mining Region of Montana</u> - This important publication on Montana Mining has not been available for over a century. Included are rare insights into the gold, copper and silver mines of the vicinity of Helena, Montana, including the Marysville Mining District, Elliston Mining District, Rimini Mining District, Helena Mining District, Clancy Mining District, Wickes Mining District, Boulder and Basin Mining Districts and the Elkhorn Mining District. Some of the topics include the early history of gold, silver and copper mining in the Helena area, insight into the geology of its mining areas, the local distribution of gold, silver and copper ores, as well their composition and how to identify them. Also included are detailed facts, history, geology and locations of over one hundred gold, silver and copper mines in the area . **8.5″ X 11″, 162 ppgs, Retail Price: $14.99**

<u>Mines and Geology of the Garnet Range of Montana</u> - This important publication on Montana Mining has not been available for over a century. Included are rare insights into the gold, copper and silver mines of the vicinity of this important mining area of Montana. Some of the topics include the early history of gold, silver and copper mining in the Garnet Mountains, insight into the geology of its mining areas, the local distribution of gold, silver and copper ores, as well their composition and how to identify them. Also included are detailed facts, history, geology and locations of numerous gold, silver and copper mines in the area . **8.5″ X 11″, 100 ppgs, Retail Price: $11.99**

<u>Mines and Geology of the Philipsburg Quadrangle of Montana</u> - This important publication on Montana Mining has not been available for over a century. Included are rare insights into the gold, copper and silver mines of the vicinity of this important mining area of Montana. Some of the topics include the early history of gold, silver and copper mining in the Philipsburg Quadrangle, insight into the geology of its mining areas, the local distribution of gold, silver and copper ores, as well their composition and how to identify them. Also included are detailed facts, history, geology and locations of over one hundred gold, silver and copper mines in the area **8.5″ X 11″, 290 ppgs, Retail Price: $24.99**

<u>Geology of the Marysville Mining District of Montana</u> - Included are rare insights into the mining geology of the Marysville Mining District. Some of the topics include the early history of gold, silver and copper mining in the area, insight into the geology of its mining areas, the local distribution of gold, silver and copper ores, as well their composition and how to identify them. Also included are detailed facts, history, geology and locations of gold, silver and copper mines in the area **8.5″ X 11″, 198 ppgs, Retail Price: $19.99**

<u>**The Geology and Mines of Northern Idaho and North Western Montana**</u>

See listing under Idaho.

Nevada Mining Books

<u>**The Bull Frog Mining District of Nevada**</u> - Unavailable since 1910, this publication was originally compiled by the United States Department of Interior. This volume also includes important insights into the geologic formations, faults and other aspects of economic geology in this Nevada mining district. Of particular interest are the fine details on many mines in the area, including their locations, histories, development and mineralization. Some of the mines featured include the National Bank Mine, Providence, Gibraltor, Tramps, Denver, Original Bullfrog, Gold Bar, Mayflower, Homestake-King and other mines and prospects. **8.5″ X 11″, 152 ppgs, Retail Price: $14.99**

<u>History of the Comstock Lode</u> - Unavailable since 1876, this publication was originally released by John Wiley & Sons. This volume also includes important insights into the famous Comstock Lode of Nevada that represented the first major silver discovery in the United States. During its spectacular run, the Comstock produced over 192 million ounces of silver and 8.2 million ounces of gold. Not only did the Comstock result in one of the largest mining rushes in history and yield immense fortunes for its owners, but it made important contributions to the development of the State of Nevada, as well as neighboring California. Included here are important details on not only the early development and history of the Comstock, but also rare early insight into its mines, ore and its geology. **8.5″ X 11″, 244 ppgs, Retail Price: $19.99**

Colorado Mining Books

<u>**Ores of The Leadville Mining District**</u> - Unavailable since 1926, this publication was originally compiled by the United States Department of Interior. This volume also includes important insights into the ores and mineralization of the Leadville Mining District in Colorado. Topics include historic ore prospecting methods, local geology, insights into ore veins and stockworks, the local trend and distribution of ore channels, reverse faults, shattered rock above replacement ore bodies, mineral enrichment in oxidized and sulphide zones and more. **8.5″ X 11″, 66 ppgs, Retail Price: $8.99**

<u>**Mining in Colorado**</u> - Unavailable since 1926, this publication was originally compiled by the United States Department of Interior. This volume also includes important insights into the mining history of Colorado from its early beginnings in the 1850's right up to the mid 1920's. Not only is Colorado's gold mining heritage included, but also its silver, copper, lead and zinc mining industry. Each mining area is treated separately, detailing the development of Colorado's mines on a county by county basis. **8.5″ X 11″, 284 ppgs, Retail Price: $19.99**

<u>Gold Mining in Gilpin County Colorado</u> - Unavailable since 1876, this publication was originally compiled by the Register Steam Printing House of Central City, Colorado. A rare glimpse at the gold mining history and early mines of Gilpin County, Colorado from their first discovery in the 1850's up to the "flush years" of the mid 1870's. Of particular interest is the history of the discovery of gold in Gilpin County and details about the men who made those first strikes. Special focus is given to the early gold mines and first mining districts of the area, many of which are not detailed in other books on Colorado's gold mining history. **8.5″ X 11″, 156 ppgs, Retail Price: $12.99**

<u>Mining in the Gold Brick Mining District of Colorado</u> - Important insights into the history of the Gold Brick Mining District, as well as its local geography and economic geology. Also included are the histories and locations of historic mines in this important Colorado Mining District, including the Cortland, Carter, Raymond, Gold Links, Sacramento, Bassick, Sandy Hook, Chronicle, Grand Prize, Chloride, Granite Mountain, Lucille, Gray Mountain, Hilltop, Maggie Mitchell, Silver Islet, Revenue, Roosevelt, Carbonate King and others. In addition to hardrock mining, are also included are details on gold placer mining in this portion of Colorado. **8.5″ X 11″, 140 ppgs, Retail Price: $12.99**

Washington Mining Books

<u>**The Republic Mining District of Washington**</u> - Unavailable since 1910, this important publication was originally published by the Washington Geologic Survey and has been unavailable for a century. Topics include the geology, rock formations and the formation of ore deposits in this important mining area of Washington State. Also included are hard to find details on the geology, history and locations of dozens of mines in the area. Some of the mines featured include the New Republic Mine, Ben Hur, Morning Glory, the South Republic Mine, Quilp, Surprise, Black Tail, Lone Pine, San Poil, Mountain Lion, Tom Thumb, Elcaliph and many others. **8.5″ X 11″, 94 ppgs, Retail Price: $10.99**

The Myers Creek and Nighthawk Mining Districts of Washington - Unavailable since 1911, this important publication was originally published by the Washington Geologic Survey and has been unavailable for a century. Topics include the geology, rock formations and the formation of ore deposits in these important mining areas of Washington State. Also included are hard to find details on the geology, history and locations of dozens of mines in the area. Some of the mines featured include the Grant Mine, Monterey, Nip and Tuck, Myers Creek, Number Nine, Neutral, Rainbow, Aztec, Crystal Butte, Apex, Butcher Boy, Molson, Mad River, Olentangy, Delate, Kelsey, Golden Chariot, Okanogan, Ohio, Forty-Ninth Parallel, Nighthawk, Favorite, Little Chopaka, Summit, Number One, California, Peerless, Caaba, Prize Group, Ruby, Mountain Sheep, Golden Zone, Rich Bar, Similkameen, Kimberly, Triune, Hiawatha, Trinity, Hornsilver, Maquae, Bellevue, Bullfrog, Palmer Lake, Ivanhoe, Copper World and many others.
 8.5" X 11", 136 ppgs, **Retail Price: $12.99**

The Blewett Mining District of Washington - Unavailable since 1911, this important publication was originally published by the Washington Geologic Survey and has been unavailable for a century. Topics include the geology, rock formations and the formation of ore deposits in this important mining area of Washington State. Also included are hard to find details on the geology, history and locations of dozens of mines in the area. Some of the mines featured include the Washington Meteor, Alta Vista, Pole Pick, Blinn, North Star, Golden Eagle, Tip Top, Wilder, Golden Guinea, Lucky Queen, Blue Bell, Prospect, Homestake, Lone Rock, Johnson, and others. 8.5" X 11", 134 ppgs, **Retail Price: $12.99**

Silver Mining In Washington - Unavailable since 1955, this important publication was originally published by the Washington Geologic Survey. Featured are the hard to find locations and details pertaining to Washington's silver mines. 8.5" X 11", 180 ppgs, **Retail Price: $15.99**

The Mines of Snohomish County Washington - Unavailable since 1942, this important publication was originally published by the Washington Geologic Survey and has been unavailable for seventy years. Featured are details on a large number of gold, silver, copper, lead and other metallic mineral mines. Included are the locations of each historic mine, along with information on the commodity produced. 8.5" X 11", 98 ppgs, **Retail Price: $10.99**

The Mines of Chelan County Washington - Unavailable since 1943, this important publication was originally published by the Washington Geologic Survey and has been unavailable for seventy years. Featured are details on a large number of gold, silver, copper, lead and other metallic mineral mines. Included are the locations of each historic mine, along with information on the commodity.   8.5" X 11", 88 ppgs, **Retail Price: $9.99**

Metal Mines of Washington - Unavailable since 1921, this important publication was originally published by the Washington Geologic Survey and has been unavailable for nearly ninety years. Widely considered a masterpiece on the Washington Mining Industry, "Metal Mines of Washington" sheds light on the important details of Washington's early mining years. Featured are details on hundreds of gold, silver, copper, lead and other metallic mineral mines. Included are hard to find details on the mineral resources of this state, as well as the locations of historic mines. Lavishly illustrated with maps and historic photos and complete with a glossary to explain any technical terms found in the text, this is one of the most important works on mining in the State of Washington. No prospector or miner should be without it if they are interested in mining in Washington.   8.5" X 11", 396 ppgs, **Retail Price: $24.99**

Gem Stones In Washington - Unavailable since 1949, this important publication was originally published by the Washington Geologic Survey and has been unavailable since first published. Included are details on where to find naturally occurring gem stones in the State of Washington, including quartz crystal, amethyst, smoky quartz, milky quartz, agates, bloodstone, carnelian, chert, flint, jasper, onyx, petrified wood, opal, fire opal, hyalite and others. 8.5" X 11", 54 ppgs, **Retail Price: $8.99**

The Covada Mining District of Washington - Unavailable since 1913, this important publication was originally published by the Washington Geologic Survey and has been unavailable for a century. Topics include the geology, rock formations and the formation of ore deposits in this important mining area of Washington State. Also included are hard to find details on the geology, history and locations of dozens of mines in the area. Some of the mines featured include the Admiral, Advance, Algonkian, Big Bug, Big Chief, Big Joker, Black Hawk, Black Tail, Black Thorn, Captain, Cherokee Strip, Colorado, Dan Patch, Dead Shot, Etta, Good Ore, Greasy Run, Great Scott, Idora, IXL, Jay Bird, Kentucky Bell, King Solomon, Laurel, Laura S, Little Jay, Meteor, Neglected, Northern Light, Old Nell, Plymouth Rock, Polaris, Quandary, Reserve, Shoo Fly, Silver Plume, Three Pines, Vernie, White Rose and dozens of others. 8.5" X 11", 114 ppgs, **Retail Price: $10.99**

The Index Mining District of Washington - Unavailable since 1912, this important publication was originally published by the Washington Geologic Survey and has been unavailable for a century. Topics include the geology, rock formations and the formation of ore deposits in this important mining area of Washington State. Also included are hard to find details on the geology, history and locations of dozens of mines in the area. Some of the mines featured include the Sunset, Non-Pareil, Ethel Consolidated, Kittaning, Merchant, Homestead, Co-operative, Lost Creek, Uncle Sam, Calumet, Florence-Rae, Bitter Creek, Index Peacock, Gunn Peak, Helena, North Star, Buckeye. Copper Bell, Red Cross and others. 8.5" X 11", 114 ppgs, **Retail Price: $11.99**

Mining & Mineral Resources of Stevens County Washington - Unavailable since 1920, this important publication was originally published by the Washington Geologic Survey and has been unavailable for a century. Topics include the geology, rock formations and the formation of ore deposits in these important mining areas of Washington State. Also included are hard to find details on the geology, history and locations of hundreds of mines in the area. **8.5" X 11", 372 ppgs, Retail Price: $24.99**

The Mines and Geology of the Loomis Quadrangle Okanogan County, Washington - Unavailable since 1972, this important publication was originally published by the Washington Geologic Survey and has been unavailable for a century. Topics include the geology, rock formations and the formation of ore deposits in this important mining area of Washington State. Also included are hard to find details on the geology, history and locations of dozens of gold, copper, silver and other mines in the area. **8.5" X 11", 150 ppgs, Retail Price: $12.99**

The Conconully Mining District of Okanogan County Washington - Unavailable since 1973, this important publication was originally published by the Washington Geologic Survey and has been unavailable for a century. Topics include the geology, rock formations and the formation of ore deposits in this important mining area of Washington State, which also includes Salmon Creek, Blue Lake and Galena. Also included are hard to find details on the geology, mining history and locations of dozens of mines in the area. Some of the mines include Arlington, Fourth of July, Sonny Boy, First Thought, Last Chance, War Eagle-Peacock, Wheeler, Mohawk, Lone Star, Woo Loo Moo Loo, Keystone, Hughes, Plant-Callahan, Johnny Boy, Leuena, Gubser, John Arthur, Tough Nut, Homestake, Key and many others **8.5" X 11", 68 ppgs, Retail Price: $8.99**

Wyoming Mining Books

Mining in the Laramie Basin of Wyoming - Unavailable since 1909, this publication was originally compiled by the United States Department of Interior. Also included are insights into the mineralization and other characteristics of this important mining region, especially in regards to coal, limestone, gypsum, bentonite clay, cement, sand, clay and copper. **8.5" X 11", 104 ppgs, Retail Price: $11.99**

New Mexico Mining Books

The Mogollon Mining District of New Mexico - Unavailable since 1927, this important publication was originally published by the US Department of Interior and has been unavailable for 80 years. Topics include the geology, rock formations and the formation of ore deposits in this important mining area in New Mexico. Of particular focus is information on the history and production of the ore deposits in this area, their form and structure, vein filling, their paragenesis, origins and ore shoots, as well as oxidation and supergene enrichment. Also included are hard to find details, including the descriptions and locations of numerous gold, silver and other types of mines, including the Eureka, Pacific, South Alpine, Great Western, Enterprise, Buffalo, Mountain View, Floride, Gold Dust, Last Chance, Deadwood, Confidence, Maud S., Deep Down, Little Fanney, Trilby, Johnson, Alberta, Comet, Golden Eagle, Cooney, Queen, the Iron Crown, Eberle, Clifton, Andrew Jackson mine, Mascot and others. **8.5" X 11", 144 ppgs, Retail Price: $12.99**

The Percha Mining District of Kingston New Mexico - Unavailable since 1883, this important publication was originally published by the Kingston Tribune and has been unavailable for over one hundred and thirty five years. Having been written during the earliest years of gold and silver mining in the Percha Mining District, unlike other books on the subject, this work offers the unique perspective of having actually been written while the early mining history of this area was still being made. In fact, the work was written so early in the development of this area that many of the notable mines in the Percha District were less than a few years old and were still being operated by their original discoverers with the same enthusiasm as when they were first located. Included are hard to find details on the very earliest gold and silver mines of this important mining district near Kingston in Sierra County, New Mexico. **8.5" X 11", 68 ppgs, Retail Price: $9.99**

East Coast Mining Books

The Gold Fields of the Southern Appalachians - Unavailable since 1895, this important publication was originally published by the US Department of Interior and has been unavailable for nearly 120 years. Topics include the geology, rock formations and the formation of ore deposits in this important mining area of the American South. Of particular focus is information on the history and statistics of the ore deposits in this area, their form and structure and veins. Also included are details on the placer gold deposits of the region. The gold fields of the Georgian Belt, Carolinian Belt and the South Mountain Mining District of North Carolina are all treated in descriptive detail. Included are hard to find details, including the descriptions and locations of numerous gold mines in Georgia, North Carolina and elsewhere in the American South. Also included are details on the gold belts of the British Maritime Provinces and the Green Mountains. **8.5" X 11", 104 ppgs, Retail Price: $9.99**

Gold Rush Tales Series

<u>Millions in Siskiyou County Gold</u> - In this first volume of the "Gold Rush Tales" series, leading mining historian and editor Kerby Jackson, introduces us to the story of how millions of dollars worth of gold was discovered in Siskiyou County during the California Gold Rush. Lavishly illustrated with photos from the 19th Century, this hard to find information was first published in 1897 and sheds important light onto the gold rush era in Siskiyou County, California and the experiences of the men who dug for the gold and actually found it. **8.5" X 11", 82 ppgs, Retail Price: $9.99**

<u>The California Rand in the Days of '49</u> - In this second volume of the "Gold Rush Tales" series, leading mining historian and editor Kerby Jackson, introduces us to four tales from the California Gold Rush. Lavishly illustrated with photos from the 19th Century, this hard to find information was first published in 1890's and includes the stories of "California's Rand", details about Chinese miners, how one early miner named Baker struck it rich and also the story of Alphonzo Bowers, who invented the first hydraulic gold dredge. **8.5" X 11", 54 ppgs, Retail Price: $9.99**

More Mining Books

<u>Prospecting and Developing A Small Mine</u> - Topics covered include the classification of varying ores, how to take a proper ore sample, the proper reduction of ore samples, alluvial sampling, how to understand geology as it is applied to prospecting and mining, prospecting procedures, methods of ore treatment, the application of drilling and blasting in a small mine and other topics that the small scale miner will find of benefit. **8.5" X 11", 112 ppgs, Retail Price: $11.99**

<u>Timbering For Small Underground Mines</u> - Topics covered include the selection of caps and posts, the treatment of mine timbers, how to install mine timbers, repairing damaged timbers, use of drift supports, headboards, squeeze sets, ore chute construction, mine cribbing, square set timbering methods, the use of steel and concrete sets and other topics that the small underground miner will find of benefit. This volume also includes twenty eight illustrations depicting the proper construction of mine timbering and support systems that greatly enhance the practical usability of the information contained in this small book. **8.5" X 11", 88 ppgs. Retail Price: $10.99**

<u>Timbering and Mining</u> - A classic mining publication on Hard Rock Mining by W.H. Storms. Unavailable since 1909, this rare publication provides an in depth look at American methods of underground mine timbering and mining methods. Topics include the selection and preservation of mine timbers, drifting and drift sets, driving in running ground, structural steel in mine workings, timbering drifts in gravel mines, timbering methods for driving shafts, positioning drill holes in shafts, timbering stations at shafts, drainage, mining large ore bodies by means of open cuts or by the "Glory Hole" system, stoping out ore in flat or low lying veins, use of the "Caving System", stoping in swelling ground, how to stope out large ore bodies, Square Set timbering on the Comstock and its modifications by California miners, the construction of ore chutes, stoping ore bodies by use of the "Block System", how to work dangerous ground, information on the "Delprat System" of stoping without mine timbers, construction and use of headframes and much more. This volume provides a reference into not only practical methods of mining and timbering that may be employed in narrow vein mining by small miners today, but also rare insights into how mines were being worked at the turn of the 19th Century. **8.5" X 11", 288 ppgs. Retail Price: $24.99**

<u>A Study of Ore Deposits For The Practical Miner</u> - Mining historian Kerby Jackson introduces us to a classic mining publication on ore deposits by J.P. Wallace. First published in 1908, it has been unavailable for over a century. Included are important insights into the properties of minerals and their identification, on the occurrence and origin of gold, on gold alloys, insights into gold bearing sulfides such as pyrites and arsenopyrites, on gold bearing vanadium, gold and silver tellurides, lead and mercury tellurides, on silver ores, platinum and iridium, mercury ores, copper ores, lead ores, zinc ores, iron ores, chromium ores, manganese ores, nickel ores, tin ores, tungsten ores and others. Also included are facts regarding rock forming minerals, their composition and occurrences, on igneous, sedimentary, metamorphic and intrusive rocks, as well as how they are geologically disturbed by dikes, flows and faults, as well as the effects of these geologic actions and why they are important to the miner. Written specifically with the common miner and prospector in mind, the book will help to unlock the earth's hidden wealth for you and is written in a simple and concise language that anyone can understand. **8.5" X 11", 366 ppgs. Retail Price: $24.99**

<u>Mine Drainage</u> - Unavailable since 1896, this rare publication provides an in depth look at American methods of underground mine drainage and mining pump systems. This volume provides a reference into not only practical methods of mining drainage that may be employed in narrow vein mining by small miners today, but also rare insights into how mines were being worked at the turn of the 19th Century. **8.5" X 11", 218 ppgs. Retail Price: $24.99**

Fire Assaying Gold, Silver and Lead Ores - Unavailable since 1907, this important publication was originally published by the Mining and Scientific Press and was designed to introduce miners and prospectors of gold, silver and lead to the art of fire assaying. Topics include the fire assaying of ores and products containing gold, silver and lead; the sampling and preparation of ore for an assay; care of the assay office, assay furnaces; crucibles and scorifiers; assay balances; metallic ores; scorification assays; cupelling; parting' crucible assays, the roasting of ores and more. This classic provides a time honored method of assaying put forward in a clear, concise and easy to understand language that will make it a benefit to even beginners. **8.5″ X 11″, 96 ppgs. Retail Price: $11.99**

Methods of Mine Timbering - Originally published in 1896, this important publication on mining engineering has not been available for nearly a century. Included are rare insights into historical methods of timbering structural support that were used in underground metal mines during the California that still have a practical application for the small scale hardrock miner of today. **8.5″ X 11″, 94 ppgs. Retail Price: $10.99**

The Enrichment of Copper Sulfide Ores - First published in 1913, it has been unavailable for over a century. Topics include the definition and types of ore enrichment, the oxidation of copper ores, the precipitation of metallic sulfides. Also included are the results of dozens of lab experiments pertaining to the enrichment of sulfide ores that will be of interest to the practical hard rock mine operator in his efforts to release the metallic bounty from his mine's ore. **8.5″ X 11″, 92 ppgs. Retail Price: $9.99**

A Study of Magmatic Sulfide Ores - Unavailable since 1914, this rare publication provides an in depth look at magmatic sulfide ores. Some of the topics included are the definition and classification of magmatic ores, descriptions of some magmatic sulfide ore deposits known at the time of publication including copper and nickel bearing pyrrohitic ore bodies, chalcopyrite-bornite deposits, pyritic deposits, magnetite-ileminite deposits, chromite deposits and magmatic iron ore deposits. Also included are details on how to recognize these types of ore deposits while prospecting for valuable hardrock minerals. **8.5″ X 11″, 138 ppgs. Retail Price: $11.99**

The Cyanide Process of Gold Recovery - Unavailable since 1894 and released under the name "The Cyanide Process: Its Practical Application and Economical Results", this rare publication provides an in depth look at the early use of cyanide leaching for gold recovery from hardrock mine ores. This volume provides a reference into the early development and use of cyanide leaching to recover gold. **8.5″ X 11″, 162 ppgs. Retail Price: $14.99**

California Gold Milling Practices - Unavailable since 1895 and released under the name "California Gold Practices", this rare publication provides an in depth look at early methods of milling used to reduce gold ores in California during the late 19th century. This volume provides a reference into the early development and use of milling equipment during the earliest years of the California Gold Rush up to the age of the Industrial Revolution. Much of the information still applies today and will be of use to small scale miners engaging in hardrock mining. **8.5″ X 11″, 104 ppgs. Retail Price: $10.99**

Leaching Gold and Silver Ores With The Plattner and Kiss Processes - Mining historian Kerby Jackson introduces us to a classic mining publication on the evaluation and examination of mines and prospects by C.H. Aaron. First published in 1881, it has been unavailable for over a century and sheds important light on the leaching of gold and silver ores with the Plattner and Kiss processes. **8.5″ X 11″, 204 ppgs. Retail Price: $15.99**

The Metallurgy of Lead and the Desilverization of Base Bullion - First published in 1896, it has been unavailable for over a century and sheds important light on the the recovery of silver from lead based ores. Some of the topics include the properties of lead and some of its compounds, lead ores such as galenite, anglesite, cerussite and others, the distribution of lead ores throughout the United States and the sampling and assaying of lead ores. Also covered is the metallurgical treatment of lead ores, as well as the desilverization of lead by the Pattinson Process and the Parkes Process. Hofman's text has long been considered one of the most important early works on the recovery of silver from lead based ores. **8.5″ X 11″, 452 ppgs. Retail Price: $29.99**

Ore Sampling For Small Scale Miners - First published in 1916, it has been unavailable for over a century and sheds important light on historic methods of ore sampling in hardrock mines. Topics include how to take correct ore samples and the conditions that affect sampling, such as their subdivision and uniformity. Particular detail is given to methods of hand sampling ore bodies by grab sample, pipe sample and coning, as well as sampling by mechanical methods. Also given are insights into the screening, drying and grinding processes to achieve the most consistent sample results and much more. **8.5″ X 11″, 124 ppgs. Retail Price: $12.99**

The Extraction of Silver, Copper and Tin from Ores - First published in 1896, it has been unavailable for over a century and sheds important light on how historic miners recovered silver, copper and tin from their mining operations. The book is split into three sections, including a discussion on the Lixiviation of Silver Ores, the mining and treatment of copper ores as practiced at Tharsis, Spain and the smelting of tin as it was practiced by metallurgists at Pulo Brani, Singapore. Also included is an overview and analysis of these historic metal recovery methods that will be of benefit to those interested in the extraction of silver, copper and tin from small mines. **8.5" X 11", 118 ppgs. Retail Price: $14.99**

The Roasting of Gold and Silver Ores - First published in 1880, it has been unavailable for over a century and sheds important light on how historic miners recovered gold and silver rom their mining operations. Topics include details on the most important silver and free milling gold ores, methods of desulphurization of ores, methods of deoxidation, the chlorination of ores, methods and details on roasting gold and silver ores, notes on furnaces and more. Also included are details on numerous methods of gold and silver recovery, including the Ottokar Hofman's Process, the Patera Process, Kiss Process, Augustin Process, Ziervogel Process and others. **8.5" X 11", 178 ppgs. Retail Price: $19.99**

The Examination of Mines and Prospects - First published in 1912, it has been unavailable for over a century and sheds important light on how to examine and evaluate hardrock mines, prospects and lode mining claims. Sections include Mining Examinations, Structural Geology, Structural Features of Ore Deposits, Primary Ores and their Distribution, Types of Primary Ore Deposits, Primary Ore Shoots, The Primary Alteration of Wall Rocks, Alterations by Surface Agencies, Residual Ores and their Distribution, Secondary Ores and Ore Shoots and Vein Outcrops. This hard to find information is a must for those who are interested in owning a mine or who already own a lode mining claim and wish to succeed at quartz mining. **8.5" X 11", 250 ppgs. Retail Price: $19.99**